Canyon

Interludes

Canyon

Between White Water & Red Rock

Interludes

PAUL W. REA

Signature Books • Salt Lake City

COVER DESIGN BY CLARKSON CREATIVE

∞ *Canyon Interludes* was printed on acid-free paper and was
composed, printed and bound in the United States.

2000   99   98   97   96   5   4   3   2   1

Library of Congress Cataloging-in-Publication Data
Rea, Paul Wesley
Canyon interludes : between white water and red rock /
by Paul W. Rea.
p.   cm.
ISBN 1-56085-054-X (pbk.)
1. Colorado Plateau—Description and travel. 2. Natural history—
Colorado Plateau. 3. Rea, Paul Wesley—Philosophy.
I. Title.
F788.R43 1996
917.91'3—dc20   96-18052
                                              CIP

Once in his life a man ought to concentrate
his mind upon the remembered earth. He ought to
give himself up to a particular landscape in his experience
to look at it from as many angles as he can,
to wonder upon it, to dwell upon it.

—N. SCOTT MOMADAY

Contents

Acknowledgements

Southern Utah University professors James Aton, Department of Language and Literature; Robert Eves, Physical Sciences; Jeff Hill, Biology; Richard Kennedy, Physical Sciences; and Al Tait, Biology. Also Brigham Young University professor Kim Harper, Department of Botany.

University of Northern Colorado professors Bill Harmon, Department of Biological Sciences; Rita Kiefer, English; Mark Leichliter, English; plus editors and critics Becky Edgerton and Jennifer Geerlings.

Colorado National Monument ranger Hank Schock, Utah Division of Wildlife biologist Jeff Grandison, Zion National Park naturalist David Rachlis, and *Deseret News* reporter Jerry Spangler.

Deborah McGee, Barbara Mogck, and Lynne Pickens, companions and commentators; Carl Putz, writer and friend; Steve Susoeff, writer and editor; Jeff Tracy, contributor and friend; Matt Vandaleur, master boatman and mentor; and Doug Rippe, iconoclast at large.

Preface

*Nowhere else on earth have I seen the possibilities for
things to proceed in such beauty towards hope for a
positive future. Southwestern landscape, the parts of it
that remain unbowed or undisturbed, seem to offer the
deepest and the most positive connections to our origins,
and a future where not all is destroyed.*

—John Nichols

The canyonlands of the American Southwest enchant
nearly everyone. Recognized as an international treasure,
the Colorado Plateau is a World Heritage Site. A land of
passionate hues, raw, brilliant, and unspoiled, this canyon
country leaves a strong impression on even the most casual
visitor.

Luminous light, wild rivers, crystalline skies, expansive
space, and gloriously exposed bedrock all intensify the
impact of this country. Primal colors prevail—bold, con-
trasting greens, browns, reds, blues. Even the infrequent
cloudy days are seldom drab and, at night, stars often blaze
down to the horizon. These canyonlands are places of
unique rock forms, lush hanging gardens, and amber twilight in the afternoon. Like distinctive Greek islands, each
canyon projects its own ambience.

This is also rugged terrain, much of it high desert that
has resisted development. There are still places where the
only prints are those of pack rats or mountain lions. Unlike
areas that are hazy or overgrown with foliage, this is great

country for seeing. Exposed rock walls, well spaced plants, and visible wildlife inhabit the vastness between earth and sky. It is not uncommon to spot a thunderstorm a hundred miles off or, in a winding slot canyon, to hear a drip minutes before it comes into view. Even the ordinary—a wind-sculpted juniper or a boldly streaked cliff—enlarge our underdeveloped capacity for wonder.

With its expansive vistas, bare walls, and deep silences, the desert enforces a degree of sensory deprivation, yet with its vibrant colors, blinding light, and extreme temperatures it also generates sensory overload. The desert invites intimate contact yet it also forbids familiarity. Because plant life is generally sparse, the plateau confronts us with otherness—stark bedrock that occasions inner change, forcing us to redefine our relation to the natural world.

Among those who ache for recharge, for fresh energy, the Southwest exerts a powerful attraction. In his essay on "New Mexico," novelist and seeker D. H. Lawrence extolled a transforming intimacy with the earth: "To come into immediate felt contact and so derive energy, power, and a dark joy. This . . . sheer naked contact, without intermediator, is the root of religion." Today Terry Tempest Williams asks similarly, "What would it be like to make love to the world?"

Amid so much barrenness, each juniper and lizard seems miraculous simply because it lives where all life seems remarkable. As William Lee Stokes observed in his *Geology of Utah*, the naked rock is fascinating because it embodies the extraordinary: "Bare rock is less common than covered rock, smooth rocks rarer than rough rocks, and light-colored rocks rarer than dark-colored ones." Because this is such an unusual environment, it requires psychological adjustments: How do we attune ourselves to so much space, so much dryness, so little green? Beyond our senses, how do we access all this?

While nature essays attempt to speak for plants and animals, for rocks, air, and water, they also render human reactions. Art critic John C. Van Dyke's impressionistic *The Desert* reveals such transformative effects, as do young Everett Ruess's heartfelt letters home in *A Vagabond for Beauty*. Uttered by a drama critic who left New York for Tucson, Joseph Wood Krutch's *Voice of the Desert* celebrates discoveries made in mid-life. And Edward Abbey's classic *Desert Solitaire* captures his way of experiencing the outdoors with abandon, even outrage. In each case the writer subjects himself to the reality of a strange landscape.

This is also a book about such resonances between earth and mind. My essays attempt to advance this tradition of the humanist evolving into a naturalist, drawn in to the desert like a moth to a sacred datura bloom. For two decades I have made pilgrimages to the Southwest, but now I savor the red rocks through the window of my study, torn between roots and wings.

For anyone, but especially for the writer, the challenge is to experience nature wide open to new perceptions. We nature writers often reveal privileged moments when we come fully alive, when the energies of a special time and place course through our being. In these intense moments our minds, hearts, and bodies embrace natural wonders, however minute, that our duller selves would miss. The mundane becomes magical, and when we respond deeply to nature we're changed, awakened to more magic in our daily lives. Most importantly, those of us who've made new connections with the natural world become more likely to treat it better.

Wilderness seekers have long affirmed the benefits of their forays: awe, amazement, healing, transcendence of the ego, spiritual purification, an expanded sense of self, a healthy humility, a rejuvenation of the more primal self. Today research corroborates nature's salutary influences: natural scenes visible from hospital windows can aid in patients' recovery. More than most landscapes, however, the profoundly beautiful, engaging canyon country is re-storative, even redemptive. How tragic that each year it faces more threats of degradation.

In tracing my growth from greenhorn to graybeard, *Canyon Interludes* deals with learning about and learning from the natural world. Greater comprehension deepens one's vision of nature. Direct experience leads us back to where we once belonged, striking deep into heart and mind, body and soul.

1.

Chaos on the Colorado

Chaos is the law of nature. Order is the dream of man.
—Henry Adams

Ignorant about what's out there, Jeff and I crave wilderness. As we pull into the Bureau of Land Management lot in Grand Junction, Colorado, Jeff quips about "the Bureau of Livestock and Mining."

"You a native, son?" drawls a rancher from his mud-splattered pickup.

"Nope, outstate agitator from Nebraska," Jeff retorts.

"How much land do you Sahara Clubbers need, anyway, just to look at?" It may take a while to feel at home out here.

We roll up the window, lock, and head inside. I ask the river ranger, a fish biologist turned paper shuffler, about this stretch of the Colorado River—especially about Rattlesnake Canyon just west of Colorado National Monument.

"Never gotten out there, but those arches are becoming better known." He hands us a crude map that shows an "unmaintained trail" climbing from the river to the arches. We decide we'd better see them before they're overrun by folks like us.

We'll float only fifteen miles, from Fruita bridge to the railroad cut. Since there's no road to this takeout point, we'll pack our gear from the river, following the cut to I-70, and then hitchhike back. This stretch of river is scenic, roadless, and, so far as we can see, without hazards.

Eager to launch our eleven-foot raft, we wedge in packs, sleeping bags, cooler, tent, and life jackets. As we each hoist an end, however, the boat buckles and flattens the snake weed. When Jeff pushes off, his shoe sticks in the mud. It's hard to break away.

Soon we're on the river. The swallows swoop, skim the flotsam on its surface, and swerve up to their nests beneath the bridge. As the parent birds bolt from their mud-daubed abodes, the peeping stops. Little rust-and-white faces fill the holes. Soon these fledglings will take their first plunge, fly or die.

We rock in the riffles, spinning with the swirls as cumulus clouds wheel over Grand Valley. This is a classic Southwest riverscape: the light brown water dances with sky blues and willow greens, then snakes between red rocks that flare into the cobalt sky.

Beside her small outboard, a Fish and Wildlife Service biologist peers into a pail. She sputters about the green sunfish, a species introduced from the Midwest, that eat the fry of native species. As voracious predators, large Colorado squawfish once reigned throughout the river basin. But the once-prolific squawfish faced threats from all directions. Locals trashed them as supposed competitors with newcomer trout, often mounted huge squawfish as trophies, and, after pitch-forking them out of the shallows, sold them by the wagon-loads for fertilizer. When squawfish fed on the channel catfish which were introduced into the local rivers, they often choked on the spines.

Though squawfish have survived three million years, adapting to drastic changes in climate, a century of human presence has nearly exterminated them. Introduced species, extermination programs, dams that interrupt flood flows, and pollution are pressing other indigenous fish toward extinction as well. In fact, river animals are disappearing much faster than terrestrial species.

"Americans may have to decide whether they want sport fish or native species in their streams," the biologist mutters as we let go, unable to resist the current any longer. Eventually evolution may replace the species lost in the present spasm of extinctions. But on any time scale, extinction is forever. Humans swim behind these now-obscure fish, hooked to a common biological destiny.

Civilization recedes as Grand Valley opens into a desert where rocks without names loom larger than language. Red sandstone bluffs tower to the left where windshields glint in Colorado National Monument; drab Bookcliffs, their spines casting shadows, mark the right horizon. As they stretch for two hundred miles, the Bookcliffs expose Mancos shale rich in fossils.

Pulling on a life jacket, I roll into the great flow. As the river overwhelms my mind, sensations flood nerves abused by urban stresses. With pulverized driftwood bobbing beside my face, I float ahead of the boat. As I stroke against the current, the Colorado exerts its power. Spitting back its spray perpetuates the illusion of resistance, though it also splatters my sunglasses.

Then a huge uprooted cottonwood looms ahead. Splashing frantic strokes, I avert the snag where the current races underneath the trunk, right through clogged branches which are notorious for netting boats and bodies. Though I've seen strainers before, with so few trees around I had assumed that the channel was clear. During runoff time a river this size rolls boulders like pebbles in a ditch; apparently it can also drag a tree. I paddle back to the raft, angry at myself.

"Why the hell didn't you call out that snag?" I sputter.

"I was busy looking for Rattlesnake," he retorts. "Why'd you get ahead of the raft?"

While I'd experienced the river, Jeff had scanned the bank for a side canyon, but few of any size appeared. Using a compass along with our map, we try to locate this bend in Horsethief Canyon, an easy place to get lost. When rustlers drove stolen horses along the river and corralled them in side drainages, lawmen seldom found the right box canyon. So it shouldn't surprise us not to know where we are.

Once oriented, we realize that we've overshot Rattlesnake Canyon. We land, deflate the raft, hide it under a bush, and hike upstream. Hauling our food and gear, we empty a canteen in the first mile. Sweat stings my eyes, sunglasses skid down my nose. A venerable cottonwood, its shade dappling ground littered with dried catkins, bleached branches, and rugged bark, finally offers us relief.

From a branch, a great horned owl beams its yellow eyes. While we tiptoe around the tree, the owl turns its broad head to track us. Since their eyes are fixed in their sockets, owls can swivel their heads a full 180 degrees. But it's their ears rather than those haunting eyes that enable owls to detect the slightest rustle. They locate prey by sound, aided by an asymmetry in their ears that allows them to hear directionally. In addition, their wing feathers eliminate turbulence in flight so owls can hunt silently. Suddenly crows dive on the owl. Though it barely blinks, the horned owl puffs its plumage and clicks its bill. The crows caw but keep their distance.

Far above the canyon rim, a bald eagle sails on the thermals. Its black wing tips finger the updrafts as its snowy head and tail show starkly against a deep-blue background. This is a rare sighting, for bald eagles remain uncommon in the Southwest. Their numbers have fallen as the result of shootings, poisoning, pollution, electrocution on power lines, and loss of river habitat. Before the eagle disappears behind a cliff, two gray-and-white redtail hawks soar up, nipping at the eagle's tail. Without ruffling a feather, the eagle spins a full three-sixty. Jeff and I stare at each other in amazement.

This airborne ballet resembles the "cartwheel display" that bald eagles do when courting—male and female hook talons and somersault, plummeting toward the ground, often spiraling thousands of feet. Packed with rods and cones, eagles' eyes are deservedly famed. Even from lofty heights they can spot a meal, tuck back their wings, and drop like feathered bolts from the blue. Other raptors may buzz eagles but know better than to linger when eagles dive.

A breeze silvers the Russian olive trees, then it dies, leaving only the hum of flies and the whine of cicadas. Are cicadas, the noisiest of insects, actually deaf? To find out, early French entomologist and nature writer Henri Fabre dragged the village cannon into his garden and fired a salvo. It didn't silence the cicadas' monotonous grinding, which proved "that cicadas are not affected by cannon fire." And that Fabre, the renowned bug watcher, didn't take himself too seriously.

Stark sunlight now drenches the naked rock. It's still too hot to cook, so we traipse down to the river and sprawl out beneath a tamarisk bush with lacy foliage and delicate lavender blooms. A muskrat glides by on the river, tail wriggling like a snake, its mouth gripping a long cattail that leaves a wide V of ripples. When its beady black eyes blink to take us in, it starts to dive but finds that if it does it will lose its morsel. Intelligence prevails over instinct.

Strange sounds haunt the dusk. From time to time a beaver or a section of riverbank goes splash. Muscles tighten. The hiss and bubble of the river are accompanied by the deep, hollow hoot of the horned owl. Plunging out of the sunset, emitting eerie shrieks, nighthawks woo mates with their acrobatics. As Jeff plays his primal flute, I wonder whether a curious coyote might bound up to check us out, as they've been known to do. Such sounds are unsettling after we've become too accustomed to quiet, enclosed bedrooms.

When magpies jeer at us from dead branches, we rise at dawn. Before long the sunlight stings our bare bodies; this will be another hot one. Perplexed by an already crumpled map, we nevertheless hope to reach the arches high up Rattlesnake Canyon before they—and we—begin to waver in the heat. As we hike along a slickrock bench parallel to the river, we reach a deep drainage. First we wedge ourselves, arms and legs splayed, into a crevice, then we skid our way down to the canyon floor.

"That chute will be fun to get back up, won't it," Jeff grunts.

We still don't know if this is Rattlesnake Canyon. Its tan sandstone walls rear about fifty feet as our footfalls echo among the concave walls. Soon dark igneous rock surfaces, ancient granite and schist gnarled into gneiss. These rocks are remnants from the ancestral mountains that underlie the more recent Uncompagre Uplift which helped form the Rocky Mountains. Because so many other layers usually overlay them, these Precambrian rocks outcrop in relatively few places. Another is where the Colorado slices through the Uplift in Westwater Canyon, not far downstream.

As we gulp more water from the canteen, a golden eagle swoops down just above our heads. First the "whuff, whuff" of its expansive wings, sounding like a grass whip, startles me; then its piercing eye stares me down. I consider it a blessing to see an eagle up so close, but I also feel forewarned. Greeting us here, alongside this angry black rock, this big, dark bird seems ominous. This is wild country, alluring but primordial.

The sun pounds mercilessly. We continue, placing our sandy boots carefully so as not to slip on dark surfaces already too hot to touch. Our boots kick some loose slabs of brick-red Chinle shale that grate as they slide. Resting right on top of this vastly older Precambrian bench, the Chinle formation is rich in iron and uranium. We're standing on a Great Unconformity, a major gap in the earth's geologic record.

At the very least, several million years of the earth's history are gone, eroded away. No strata from the entire Paleozoic era remain. When these igneous Precambrian rocks first solidified, life had scarcely begun; when they began metamorphizing, complex marine invertebrates such as trilobites had barely evolved. Yet when the record resumes as the Chinle shale was building, fish and amphibians had long since evolved and reptiles had been around for millions of years. What happened to these missing layers? One answer is that they eroded completely away, contributing to the massive deposits of sedimentary rock and salt that lie west, in what is now eastern Utah.

We rest again, foreheads dripping. Jeff is pale, so I forget about the arches. "Shouldn't have tasted those plants," he mutters, referring to his habit of sampling leaves. The dark gneiss seethes, then shimmers like a translucent veil across the bare-boned rock. While Jeff's fingernails grope for cracks in the rock, my hands grasp for juniper branches. Neither hand nor foot grips well. I skid down the slope, heels dug in, and start a mini-avalanche. The dislodged sand and shale strike bottom, echoing softly off the walls.

Beneath these contoured cliffs, we swelter in the shade. Twigs, needles, pinyon nuts, shriveled juniper berries, and dried cottonwood leaves, their netted veins bleached by the sun, litter the narrow canyon floor. Salts crust the sand. Flies buzz around a shrinking puddle. Desperate water striders shoot this way and that as polliwogs dig themselves cones with their noses. How many more days can this life-giving puddle last? At its edge, where slimy water oozes from the sunbaked mud, ants pick at stricken tadpoles. The descending, decelerating notes of the canyon wren further energize the rippling air.

A slight draft bows cattails rooted in the cracking mud. In the sunlight their long and graceful leaves glow an almost phosphorescent green. Yet they seem no more alive than the moss-streaked rock beneath the seep or the canyon wall enlivened by reflections from the pool. Above us, weak gusts rattle the dark green cottonwood leaves as blue drops of sky trickle through holes eaten by insects. Enclosed within this box canyon, I feel breathless.

Buzz. A large bumblebee whirs as it zigzags down-canyon. Whir. A canyon wren drops down to scold. The hum of insects lingers until an ash-throated flycatcher, its pastel yellow breast gleaming, snaps a twig. It shoots out to hunt a bug, feathers flashing, and grabs it with a crack. All this commotion awakens Jeff.

"We're being invaded by birds and bugs," I inform him. He rubs his eyes. "In this dream I hiked through giant tunnels cut by ancient streams, looking for an arch. We never found it, but we sure ran ourselves out of water. These canyons are seductive, always enticing you around just one more bend."

As Jeff completes his labyrinthine reveries, sunlight creeps ever closer, puddling our cheddar into orange dye and gum. In the shade I can feel the glare from sunbeams that strafe the sand a yard away. The walls seem to melt inward.

Jeff is too sick to hike back; we'll spend the night here. After some difficulty climbing the chute, I reach our first camp to retrieve the barest essentials: stove, soup, water, and Jeff's bag. At least Jeff can stay warm and regain the water he's lost.

Crows gather under the now-leaden skies. Dozens of them swirl around a cottonwood as though a vast net with thousands of black knots had dropped down, snagging on its branches, until every twig ended in a knot. This must be their roost for the night. Odd. Has this gaggle strayed from the orchards in Grand Valley? Shadows pool beneath a blanched juniper that stands gaunt against the sky.

Rather than sleep on damp sand, we straggle toward the dark rock where at least the sand is dry. After we share the soup, I dice cattail roots, washing them down with precious water. Tonight we munch muskrat chow, forlorn and far from home.

The black rock radiates heat that this time draws me closer. To purge self-pity and to embrace this bedrock, I recall Robinson Jeffers's celebration of "Oh, Lovely Rock," of "this fate going on outside our fates." Yes, this Precambrian rock has its own fate. Today grains that have never seen life are seeing light for the first time in over a billion years. My palm rubs this ancient bedrock not just for warmth but to connect with its unfathomable past.

Yet I can't respond fully. With so much destruction and death going on today, perhaps identifying with nature is like visiting a dying friend. Some people fear growing too close, while others seek to connect with the friend before he or she changes utterly. Tonight I'm in the former, fearful place, though I long for a deeper connection.

During the night I wriggle first into the warm sand, then against Jeff's bag. I wake at daylight, cold. A dawn breeze sighs through the cottonwood, then rattles its leaves as it expires.

Jeff's breath smells rancid, but he looks less pale. Once he feels strong enough, we set off for our real campsite. Rather than climbing the chute, which requires strength he may not possess, we follow this drainage to the Colorado and slip along its bank. But soon an unforeseen problem arises. Cliffs dictate that we won't skirt the river's edge. We'll have to wade around a point, with me in the lead so Jeff won't get swept off his feet. It looks easy enough.

The river, however, doesn't care about us, and wading becomes a blind man's bluff. The current draws me into chest-deep water. I bounce on tiptoes, overstuffed daypack in one hand and camera in another, looking over my shoulder for Jeff, wondering if I'll get swept away. We stagger out like drunken hippos. Jeff remarks on going from hot to cold: "Either can do you in if you aren't careful." At camp, still chilled, Jeff rubs himself dry in a sleeping bag.

A horned toad snaps up the ants we've attracted. This

fellow is lucky, for all this heat will aid his digestion and he'll bask on a rock after we're gone. Tomorrow, buried six inches beneath the sand, he'll sleep off this feast. As we watch, we wait for energy that never comes; then we shoulder our packs.

We're relieved to find that downstream a mile our raft is undisturbed; soon it's pitching with the riffles again. The tan-and-grayish cliffs expose Morrison formation from the Jurassic age, roughly 150 million years ago. The Morrison, a layer rich in fossils, contains petrified trees that grew a hundred and fifty feet tall, just above the reach of Apatosaurus, the herbivore that grew seventy feet long. Morrison sandstones and shales have yielded some of the earth's finest fossilized bones both at Dinosaur National Monument and also near Price, Utah. Because it also contains concentrated uranium oxide, the Morrison formation has received a lot of attention.

It would be difficult to miss the railroad, but we hug the bank just to be sure. Ahead the yellow and gray bluffs of Horsethief Canyon give way to the red sandstones of Ruby Canyon. On this site government engineers once hatched plans for a dam. Years later, citing the need to preserve habitat for endangered species and to designate this spectacular stretch of river as Wild and Scenic, the Bureau of Land Management wisely recommended against the dam.

While the river licks the mud, we spot a black roadbed rippling in the sun, then the glint of shiny rails. The acrid smell of creosote-coated crossties pierces our nostrils. Here a clear Salt Creek slips into the Colorado, churning the clouds of silt. We land and flatten the raft, its valves hissing in protest. As we head down the tracks, leaving the raft for a return haul, we encounter another obstacle: a single-track tunnel. There really isn't any good way around it: cliffs drop down to the river.

Once into the dark tube, danger strikes. We hear a roar, then see the blinding beam of a locomotive hurtling toward us. Instinctively we tear off our packs and dash for daylight, eating the grit of railroad slag as we slide on cinders, gasping expletives. The fast freight thunders and screeches by. Panting, shaken, we stare at each other in disbelief. Jeff looks pale again.

"There's not going to be another of those suckers right on the tail of that one, so we'd better head through," Jeff gulps.

"OK, if you feel up to it." This tunnel, as it turns out, runs for nearly a hundred yards. We each ponder where we'd be if the train had caught us just a hundred feet farther in.

In searing sunlight the track knifes straight along Salt Wash, a strip of greenery. As we hit the greasy ties with long strides, my boot nearly tromps on a large snake stretching from rail to rail. Despite the momentum added by a heavy pack, I backpedal in midair, land on all fours, hands gouged again by the slag. Rattles quiver. As it wriggles up the cutbank like water flowing uphill, this reptile stirs fear in the dungeons of my mind. I've always liked snakes, but now I want to strike this symbol of hostility. Captain Ahab's hatred of a whale suddenly makes perverse sense. We humans resent that we're overpowered, that nature remains indifferent or even harbors malice. All this exposure to primordial rocks, to danger and even loss of control, has activated my instinctual, reptilian brain.

Soon traffic rumbles overhead on the I-70 bridge. Remarkably, Jeff feels better. He pulls on his running shoes, climbs the embankment, and plays a hobbling jogger with the turned ankle. When the second car stops, Jeff waves in triumph while I return for the raft. Near the location of the snake I tread lightly, stick in hand, but the rattler has slithered away. My focus on the serpent has allowed me to ignore an obvious question: is the raft really worth the risk?

Faced again with the black hole, I stop to listen for a roar or a whistle. Nothing. I race through, numbed to the memory it awakens. Raft and oars crammed into my backpack, I slouch forward like a *sherpa* and then begin the return trek. At the tunnel the only sounds this time are the clanking of rails expanding in the heat and the cluck of redwing blackbirds swinging on cattails. Trucks whine in the distance. As the tar and creosote fry my feet, my tepid canteen tastes of aluminum. When I trudge in, bowed under my load, I see my car sitting with its doors open. Grizzled as an old juniper, Jeff hoists an iced drink. He looks rejuvenated.

These bedrock experiences were hardly what I'd expected from a float trip and hike to the rainbows in stone. Wilderness is supposed to leave you refreshed and relaxed, but I feel battered. When I reached out to the canyon country, it slapped me down. For too long I've lived apart from nature, reading about it, forgetful of its raw power. People who live outdoors, close to it every day, don't entertain such romantic longings.

Interlude

Racing to Angel Arch

An air-conditioned jeep would allow us to see Angel Arch; why not succumb to temptation? We pooled our money and found that we had enough to rent a jeep for two hours. So off we went like tourists intent on "getting there" for a look and getting back quickly.

Soon we were barreling back down Salt Wash, slapped by

willow branches. At Peekaboo Spring we struck a broad puddle, splattering the windshield. A polliwog wriggled on the wiper. After each willow thicket I sped up until I spotted where oversized tires had rutted the brush; then I braked hard. The challenge was to cover the winding, open stretches as fast as possible, kicking up rooster tails of sand. Like a kid with a new toy.

Oblivious to parked jeeps, we first overran Angel Arch, then spun into an arc that sent sand flying until we found the main tracks again. At this backcountry lot, many treads had stamped out sand castles with crumbled ramparts. Backpackers glared as our jeep's fan droned on. I scrambled straight up to a picture point, but the light was lousy. My shot wouldn't be a postcard, so I bounded back down toward the jeep, eyes on my next landing. Irked by the rush, my companion hesitated to climb back in.

"Would you drive a little slower now?"

I looked at my watch. "Yeah, but we gotta make time so that rental guy doesn't charge us for three hours."

I drove slower going back and even stopped at Paul Bunyan's Potty, a giant hole in a sandstone overhang. This time I took longer to compose my shot of an arch. This also annoyed my friend.

"Are we here for this country or just for its freakish forms?"

"I don't know," I mumbled.

Despite the twenty-odd miles of gorgeous red rock canyons, I was nagged by emptiness. My tunnel focus on a destination had cost me more than I'd reckoned: I came away with lousy photos and a credit card slip. I had to face the fact that this was no way to see the natural world, and that speed was a problem.

Scott Slovic, a critic interested in responses to nature, even became disenchanted with jogging in nature. "Running was a form of insulation," he came to realize, "a self-imposed dream state, the rapid motion . . . and hyperventilation combining to blind me to my surroundings."

If jogs can be too insulating and too fast, then what about jeeps, or even bikes? Jeeping has become so popular around Moab, Utah, that the desert has become something of a drive-up park. By the 1990s some of the jeep routes were closed—starting with Salt Wash, where maniacal desert rats like me had damaged stream bottoms. Even mountain bikers often move too fast to feel a part of the environs. Riders whir right by a creamy sand verbena or, seduced by aggressive treads and downhill thrills, whizz right over a pincushion cactus in bloom.

Restrictions on wheeled vehicles help preserve the park, but they also protect us from our insatiable high-speed addictions. We've got to slow down.

2.

Blown Away in Fantasyland

Alone on the open desert, I have made up songs of wild,
poignant rejoicing and transcendent melancholy. . . . loved
the red rocks, the twisted trees, the red sand blowing in the
wind, the slow, sunny clouds crossing the sky, the shafts of
moonlight on my bed at night. . . . at one with the world.
. . . I have exulted in my play. I have really lived.
—Everett Ruess, letter of April 18, 1931

It's one of those cool and clear spring days after a rain, with pink and lavender wisps in the distance. Amid rock, sand, and sky, the ants are rebuilding their mounds. In May this slick-rock country can sizzle, so I smear on sun block. Since there's no reliable water here in the Needles District of Canyonlands National Park, four canteens swell my pack. Not enough, but they'll do, possibly supplemented by short-lived puddles.

Switchbacks ascend from Elephant Hill and head toward the Needles and Chesler Park. Here I can walk by feel. If a boot strays into soft sand, I know I'm off the track. Freed from their directional duties, my eyes can roam. Expanding rhythmically, my chest draws in the spicy aroma of a cliffrose. After snows and rains trickle down this bare sandstone, cliffroses explode into creamy blooms nearly covering the bush. Among the sturdy shrubs that cling to rock walls, none better typifies springtime in these canyons.

Cumulus clouds now bulge pearly gray on their bellies. Ravens circle overhead, their husky clucks bouncing off the sheer rock. A coyote lopes by, trotting sideways to keep an eye on me, then gallops beneath the ravens, glancing up at them. Animals watch each other, especially when the presence of one species implies a meal for the other.

Striped and rounded rocks, examples of the Cutler and Cedar Mesa monoliths studding the Needles and Standing Rocks areas, gleam against a mare's-tail sky. Unlike most sandstones of the Southwest, here there's less of the windswept cross-bedding and more of the flat stratification that indicates sedimentation under water. The colors—beiges, pinks, creams, grays—resonate in bands. The redder layers come from river deposits originating on the rising flank of the ancestral Rocky Mountains, the lighter sandstone bands from submerged sandbars and coastal dunes along shifting shores.

Over time the borders between the bands have blurred. The ensuing eons plus the permeability of these sandstones have allowed ancient pigments to blend like melting Neapolitan ice cream, as Ed Abbey remarked. Embodied by this lack of regular stratification and clear boundaries, this unpredictability epitomizes the incessant change that characterized the Permian period, 250 million years ago.

The trail surmounts a great whale's back of slickrock, drops in and out of drainages, and heads through troughs— wide cracks that become narrow enough to snag the ends of my rolled pad. Yet despite the exertion of a three-mile hike under a heavy pack, I need a turtleneck. Odd. On a typical late spring day, a packer would have downed a quart of water on this stretch just to cool off.

As I trudge up the ridge, gravity and wind stop me mid-stride, throwing me off balance. This slit in the rock is a hellacious wind tunnel. As my thighs tremble from the strain, my hat sails down into the amphitheater. Clouds race in to heighten the grays of the towering boulders, to blur the line between earth and sky. Throwing down my pack, I dig for a flannel shirt.

As I quite literally push on, buffeted by the wind, Chesler Park sandblasts my face. Even glacier glasses with side panels don't keep out the grit that sticks to the fogged lenses. The dust makes it difficult either to see the trail or to avert sharp yuccas. A huge conning-tower rock affords no shelter. Like chop along a breakwall, the wind whips from all directions. Glimpsing a haven through fogged and flecked sunglasses, I grope toward a huge slab with a crack behind it. Refuge. Spring in these canyons is a flirt; first she teases you into taking your shirt off, then she blasts you with sand—or even snow.

I yank on jeans right over my shorts, then huddle behind the slab, already out of extra clothing. Anticipating warm, dry weather, I haven't packed a raincoat. Although I knew that temperatures in southeastern Utah often fluctuate fifty degrees in a day, my heart yearned for spring. In a protected back yard, it's dandy to say, "I love the wind—it's a force in nature that humans haven't begun to control." But out here vulnerability to such an uncontrollable force frightens me.

Exposure like this probably occasioned the "wilderness shock" experienced by early settlers and pioneers.

Just when the sandstorm lets up, sleet swirls right into my shelter. I've got to act before hypothermia freezes me solid. My first aid kit contains a space blanket, but this thin cover flaps wildly in the gusts. Now what? Wool socks serve as mittens, and a sleeping bag becomes a makeshift tent. Though my bag now hoods my hunched body, sharp drafts still spike my legs. I shrivel further into a ball and breathe into my socks-as-gloves. As exhalation warms this burrow, I drift into a hypnotic state, beyond any sense of time and place, becoming my breath. The hail pecks on fabric to become my only awareness. My eyelids pulse a rust red, melding iron in blood with iron in rock.

This hibernation experience has felt wondrously child-like, but is there also something childish about my response? To most children, nature means sunshine, goldfish, dime-store turtles, puppies, kittens, a protected hollow in the shrubbery or a tree fort. Winter means snowmen and snow-balls on sunny days. When it gets dark or storms, kids scamper indoors. Adults, on the other hand, know rationally not to expect shelter in a desert. But when an area is so inspiring, so picturesque as this, it's difficult not to feel some shock when I experience the raw forces which fashion it.

The sleet stops and the wind slows. What time is it? What a silly question. How arbitrary to divide the day into 1,440 units, much as we fragment so many other things, including ourselves. Glancing at the left wrist is one of those habits, only one among many, that lessens the outdoor experience by recalling human constructs. Yet who can claim that he or she has not brought societal stuff into the wilds?

As I arise, sand in my brows, a strange world appears.

Even allowing for my impaired vision and altered mental state, the light seems weird, the air looks orange, just short of the "veritable sand fog" that early desert writer John Van Dyke saw after a sandstorm. Tinted snow dusts the grasses and crests the mushroom-topped monoliths, further intensifying their reds. Low clouds render the lighter bands in the rocks even more pastel. Colors intermingle. Where millions of grains of silica have pinged millions of others, the rocks gleam after their latest sandblasting.

It takes eons for sand, after countless collisions, to round down to a fine particle. Once broken off from quartz crystals, sand grains recycle through the ages. Silica sluffed off ancient granite may scuff through countless collisions over two billion years. As a result, a granule's roundness offers one clue to its age. So while the slickrock I'm treading on is a quarter of a billion years old, the grains composing it are infinitely older.

The Needles behind Chesler Park project a skyline of spires that tower four hundred feet above a lagoon of pea-green grass. During the nineteenth century huge herds grazed these canyons, nearly wiping out the native tall grasses. Starting in the 1880s, cowboys drove mile-long streams of cattle either southeast to summer pastures in the Abajo Mountains or northward through the Moab valley to the railhead at Cisco. Moab was a rough cowboy town and rustlers hid in the canyons, as did other outlaws. Around 1890 Robber's Roost became the hideout for Butch Cassidy and the Wild Bunch, who knew this convoluted landscape as well as anyone.

If the ravages of overgrazing were not enough, much of this incredible area was almost lost beneath a dam. In 1961 the new Secretary of the Interior, Stuart Udall, flew over these canyons to inspect the storage potential of Glen Canyon

Dam, then under construction. Astonished, he exclaimed, "God, that's a national park out there." Seeing is indeed beholding. Project Lighthawk, run by volunteer pilots to promote conservation, now flies government officials over areas that are at risk.

Spectacular as it is, Chesler Park seems diminished from when its stirrup-high grasses waved in the breeze. As Aldo Leopold, the founder of environmental ethics, observed, "One of the penalties of an ecological education is that one lives alone in a world of wounds." The establishment of the park in 1964 promised a gradual elimination of grazing so that the native blue grama, galleta, and rice grasses could come back. However, cheatgrass, Russian thistle, and other invader species have still not yielded. Nor has trespass grazing ceased. A 1993 Department of Interior report indicated that while most of this land is recovering from past abuses, cattle still enter because the park is not entirely fenced.

The destruction wrought by livestock runs deep, well beneath the surface. The microbiotic or cryptogamic soil, a dark crust of living earth on the Colorado Plateau, is biologically crucial. Park ecologists have found that native grasses depend on this miraculous mix of lichens, mosses, algae, and fungi whose interrelationships are concealed beneath its bumpy surface. Its spongy texture absorbs moisture and binds the sand to prevent erosion. Where it's been disturbed, the crust makes a slow comeback. Often it gets trampled again by two-hoofed critters who also churn up the sand.

Although desert flora may seem as if it can survive anything, in fact it's slow to restore. Brush flattened in the 1950s has still not grown over the jeep tracks. Since the Needles area has long been a popular backcountry destination, hikers are told to stay on trails. Although they wail about the damage done by cows, hikers often ignore their own impacts.

Chesler Park beckons, its writhing grasses as green as luna moths, its pinnacles tawny as the rumps of desert bighorn sheep. Huge fins rise from the meadow where low grasses still bend in the gusts. Seed eaters such as kangaroo rats and black-throated sparrows frequent these sparse meadows. Horned larks huddle and face the wind, their breast feathers puffed or flattened. Since they live in such open places and return to where they were hatched, these larks have little choice but to face spring sandstorms. Before today I seldom considered how many birds and bugs freeze to death in storms.

Around the giant dome in this enclosure trots a mule deer, nostrils puffing steam and whiskers quivering. Somehow, despite the cold, a fly buzzes above his tensed-up back. Then he bounds with great regularity, feet together, his black tail bobbing. Since he has recently shed his antlers, he moves with even greater grace.

When clouds engulf the sun, once again I seek a haven. Shouldering my pack, I explore alcoves and amphitheaters along the trail toward Devil's Kitchen and Needles Outpost, watching the humped slickrock undulate from point to bay. Some alcoves offer better protection than others. Finally I settle for a bay where a gnarled old juniper provides an additional windbreak.

This haven is a desert garden. Fremont barberry with holly-like leaves and desert four o'clocks, their magenta trumpets waving in the wind, both landscape the scene. A patch of golden mule's-ear sunflowers glows nearby, swirled as though painted by Van Gogh. A deep scarlet paintbrush flares inside a gray sagebrush. All these plants are well spaced, implying the open simplicity of the desert.

This cozy spot is mine for the evening, complete with its rock-enclosed yard, a private footpath, and a sheltered kitchen. After exposure to raw elements, my mind seeks security through order. One part of me craves wilderness, another fears it. In response to my dread of relinquishing control, or my need to make the strange seem familiar, my tendency is to project human forms and colors upon nature, often through names.

While I hunch over the stove, skies darken and begin to sprinkle the sand. When the flame flutters too badly to cook, I retreat to my tent where my breath clouds the greenish light. As the stakes become loosened by the gusts, the tent flaps like a plastic bag on a bush. When I venture forth, hoping to cook quickly between sandstorms, haste makes the sauce boil over. I grab tongs but the pan, glued with cheese, sticks to a stove that's blasting like a blowtorch. I reach for the nearest branch to hold down the stove. When the old branch snaps, my Deluxe Noodles With Zesty Parmesan Sauce plops on the sand.

I howl, but the wilderness engulfs my cry. As canyon writer Ray Wheeler notes, this experience leads to "the sudden diminution of my ego—of my sense of importance—against the scale of the landscape, the planet, the cosmos." We hope that nature will humble us, but when it does we resent the deflation of our egos.

The rocks seem to mock my hunger. One hoodoo resembles a chef's white hat, another a voluptuously-rounded clover leaf roll. I maraud my own campsite, tearing into a bag of almonds and pinion nuts. Honey and crackers stick in my whiskers. Solitude allows me to forget the rules of polite society and even enables me to become an animal. To reconnect with nature, I let the beast off the leash. Activating the wilds inside links me to the wilds outside, as well as to a deeper self. Wilderness is a place to vent passions—fear, anger, even hate—that the social world won't accept. When we allow it to, wilderness fosters re-creation in the best root sense.

Once feeding time is over, I bound down the slickrock and spider up to its rounded landing. Spidering, really scuttling around on all fours, face up, not only gets the rock scrambler to otherwise inaccessible places, but also offers a way to know the rock more intimately. As I spider up or down increasingly steep inclines, my fingers explore the sugary surface of the sandstone, groping for niches or ridges. My fingernails grip and grind like claws as I skid down a chute.

Something stops me mid-breath. Silence steals my breath as the vast landscape seems to listen. Is this fairyland real or just painted in pastels? Far below, gloaming in the twilight, stands a sand castle. High ramparts, topped by observation towers, rise above the straight boulevard leading through gates. Beyond them rises a medieval hill town, its buildings spaced regularly. Farther below are houses with mushroom caps for roofs. Some lots remain vacant, much like those at Pompeii.

As explorers observed, this country jars our sense of proportion and human scale. Its vastness both exhilarates and distorts. In 1859 explorer Captain J. N. Macomb imagined the Needles as a Manhattan bristling with spires twice as high as Trinity Church. His journal keeper saw "battlemented towers of colossal but often beautiful proportions, closely resembling the elaborate structures of art." With so much naked rock, much of it weathered into surreal shapes, the mind sees spires and pinnacles in familiar terms. We humans can't commune with this much bedrock otherness, so our imagination relieves our sense of alienation. But wait, why pathologize a lovely fantasy? The self-examined life can become the opiate of the introspective.

Back at the campsite one wood rat gorges itself on my disaster of a dinner. Another munches on my crackers. One jumps when it hears my footfall but soon returns to the banquet. Sand is no deterrent to them; in fact, it's a feature of their high-silicon diet. I scan the site for any shiny possessions that my bushy-tailed friends might carry off. These cliff dwellers will pick up anything they can heft. The Needles District boasts several types of wood rats, with more kinds living in close proximity here than anywhere else in the world.

One well-fed rat scampers up the cliff into a blowhole that's filled with dung from countless generations. Individual wood rats live alone, but their dens, fortified by the feces and urine that cement leaves, twigs, and sticks, provide good protection. Preserved by the desert's dryness, some nests date back thousands of years. Because of their great antiquity, such dens provide information about changing climates and vegetation well before the last Ice Age.

Old pack rat nests also preserve fragments of plants and animals encased, amber-like, in crystallized urine. This led to some amusing misunderstandings among hungry pioneers: "We found balls of a glistening substance looking like pieces of variegated candy sticking together," observed a prospector headed for California in 1849. This fellow assumed that this fossilized rat excrement "was evidently food of some sort, and we found it sweet but sickish." Though I'm famished right now, I'll gladly share the secret of this natural rock candy with desert rats of all sorts.

At home in the soft sand, I recline on the slickrock to watch the grainy twilight. In the grand distance, far beyond the Henry Mountains sixty miles away, the sunset highlights the horizon. A new moon sails the sky like a boat in the blue. A passenger jet blinks into the sunset. I recall myself at thirty thousand feet, nose pressed to the bubbly glass, longing to explore the wrinkled landscape below. Once the jet's rumble expires, the silence of the desert almost seems to scream in protest.

I arise early and explore the imaginary castles and towns. With only one canteen I stride easily along the peninsulas of slickrock, jumping the deep, straight joints that separate them. Amid all this curvature, the grabens—shallow, sheer-walled valleys—also seem both welcome and incongruous. These two straight features both reflect the geology here.

To comprehend these complex processes, one needs to back up over three hundred million years. Early in the Pennsylvanian era, as the first dinosaurs were hatching, the central part of what is now the Colorado Plateau began to sag, filling with seawater. Over many millennia, as this basin settled and more seawater first entered and then receded, massive deposits of salts accumulated. At its maximum extent the Paradox Basin extended from northern New Mexico to central Utah, covering much of this area with thousands of feet of salt, gypsum, and shale.

About three hundred million years ago, as sediments began to fill the basin, the salts became viscous enough to ooze upward. They often, in the words of geologist Donald Baars, arose as "intrusive salt walls that pierced all overlying strata in an attempt to follow the path of least resistance." Under great pressures this salt continued to bulge for nearly 150 million years, often arching into domes within the rock. Later the groundwater dissolved the underlying salt. As cavities in the rock enlarged, the surface layers often collapsed.

The result was the parallel faulting that formed long "salt valleys," one of which encloses Moab, Utah.

Whereas settling formed the grabens, fracturing explains the joints. During the middle Permian period, when mammal-like reptiles were evolving 250 million years ago, deep sand dunes began mounding in what is now southeast Utah. Later, during the Jurassic age of dinosaurs, limestones began to build on the sandstones. Much later, about sixty million years ago, the Monument Upwarp lifted what is now southeast Utah, northern Arizona, and northeastern New Mexico. As the Upwarp heaved, the overlying Cedar Mesa sandstone fractured into great blocks.

Because the Upwarp sloped toward the junction of the Colorado and the Green rivers, when subterranean streams flowed down this slope, they dissolved the salts. Thus undermined, some of the huge blocks sagged while others subsided. The resulting great cracks, widened further when the flexible salt deposits slid toward the Colorado River, furnished the rough cuts. Rain, ice, wind, sand, and gravity are adding finishing touches to the more resistant spires, needles, and fins. These are well worn lands, eroded down to resistant remnants.

Beneath this wavy gray slickrock, I find a route down to the city of striped spires. A roadrunner perched on a bleached juniper suns its striped feathers. Below it a whiptail lizard hunts on a sunny rock, thrashing its tail as it cocks its head at a fly. Suddenly predator becomes prey. A roadrunner grabs its meal, then springs four feet straight up into a juniper. The whiptail hangs limp in its beak. Though they're not good fliers and therefore don't migrate, roadrunners can sprint at fifteen miles an hour. Most large hawks lack the ground speed and agility to capture these lizards, but roadrunners have made them a speciality.

Last night's boulevard leads toward the castle. In full daylight these giant monoliths look less like human constructs, though their straight cleavages remain astounding in any light. I wander among the towering rocks, seeing and hearing no one. The Joint Trail, one of the many crevices in the sandstone that encircles Chesler Park, runs straight like a trench. Keeping my topographic map handy, I follow washes and jeep trails. Clouds keep temperatures in the sixties, conserving my water. Canyon upon canyon winds on and on in a riot of erosion. As Ebeneezer Bryce remarked, this would be "a helluva place to lose a cow."

My boots crunch on the pebbles of the drainage, allowing navigation by ear, while my eyes lead me around just one more corner. As rock needles reach toward the clouds, I try to note those striped with rust-and-cream bandings, but there are hundreds of such rocks in just this canyon alone. This proves about as useful as leaving a trail of crumbs.

Before long I'm lost in a labyrinth. Which way now? Following my own footprints doesn't help much, for too many boots have tramped these washes. But if so many people have trodden here, how come I've gotten lost? A swig from my canteen washes down the last of my trail mix. Getting lost may signify something, such as the importance of watching for truly distinctive landmarks. As I pour over my map again, compass in hand, I hear voices.

Surprise. Well-groomed jeepers offer beer and sandwiches. While they've four-wheeled legally to arrive here, they do nevertheless dispel the backcountry magic. For me, at least, such interactions trigger a reversion to societal consciousness. If "Have a nice day" rings hollow on the sidewalk, it sounds much worse on a backcountry trail.

This said, however, I'm very glad to find my bearings, to

swig the cold beer, and to scavenge the food. Now that I know my whereabouts, I return to camp before I get lost again. The question is, should I stay another night, skimping on food and water, or hike out before dark? Another night of magic is worth the discomfort, I figure. Dinner consists of an elegant hors d'oeuvres course of trail mix, followed by a main entrée of peanut butter, enhanced by some vintage *eau de canteen*. I pass on the local rock candy for dessert.

At dusk, hoping to see deer in Chesler Park, I crouch behind a rock to wait, downwind. As the light wanes toward a Maxwell Parrish blue, the air turns grainy again. My eyes strain. No deer. I rub my arms quietly and wait, shifting from foot to foot. The sky has now become a deep blue, well on its way toward indigo. No deer. Strange, for Chesler Park offers the best forage for miles around. In the cool twilight air I stride back to my tent, warming up and making no effort to lighten my footfalls.

Then movement strikes the corner of my eye. Two big, broad ears, a rack of antlers, and then many more ears loom in the dusk, not forty feet away. These deer are stalking me, pacing silently back and forth, ears twitching with the slightest noise I make. So curious. The buck weaves his way in closer, watching me all the time, then lifts a hoof to scratch his flank. What do these deer want? This is a popular backcountry area; they probably expect handouts. For minutes on end I watch these visitors, my gaze panning from dark eyes to dark eyes. Suddenly feeling the chill, I ease open the zipper of my tent and slip inside.

Hooves thud on sand and grind on rock, followed by snorts of frustration like those of a dog that sticks its head into an empty dish. Something obscures the stars that were penetrating the pinholes of my tent. A large animal is standing over me, its sharp hooves separated from my face by only a tissue of nylon. Pans clatter. My stomach gurgles. I peer out the window. One doe chomps on my visor, a last line of defense against the desert sun. These beasts actually seem aggressive. "They're only deer," whispers my rational mind, yet my heart races. As the deer move on, my head eases onto the pillow, ears twitching.

To conserve water I rise before dawn to hike while it's cool. Venus, the brightest star in the sky, dances in the east. Called "phosphorus" by the Greeks, the Goddess of Love beams a dozen times more brightly than Sirius, her closest rival. Yearnings for love put a bounce in my step as I complete my desert sojourn. On the other side of the Rockies, my beloved Barbara awaits my return.

On the trail back, this time without the wind in my face, the scenery is glorious. Fins and pinnacles tower above a great natural amphitheater where the vistas are sensational. Just beyond my boot, a shiny black wasp drags a brown spider, its legs still limp, toward its nest. Below lie the white-capped giant mushroom rocks of Elephant Canyon. In the mid-distance, the sheer cliffs of Needles Overlook soar far above the slickrock pinnacles, their edges sharp in the crystalline air. Behind them, nearly sixty miles away, rise the snow-covered La Sal peaks. Two days ago, in my rush to avoid the wind and sand, I missed all this when I kept my head down except to watch my airborne hat.

Descending into Elephant Canyon, I see more that I missed. Anasazi petroglyphs and pictographs stand out at a confluence of canyons. On the right side are four delicately-lined hand prints. In this location, a natural route out of the drainage, one imagines that this rock art functioned (at least in part) as signposts. If so, the directions they indicated remain well beyond my ken.

The canyon floor is paved by hard gray rock. Embedded in it is a fossil tooth within the Elephant Canyon formation, which was deposited roughly 280 million years ago. Here in Canyonlands the rock from this period runs up to a thousand feet thick but outcrops in only a few places. Since my canteen's nearly empty, I scan the drainage for water pockets, but there's no water here, at least none that I can access.

In the chute formed by a joint, vine-like roots stretch like conduits through a steam tunnel for nearly a hundred feet. An intriguing tree unfolds its serrated leaves. Its branches hang clusters of flattened but leafy hop-like bladders, each containing a seed. The ranger later informs me that this is a rare Western hop hornbeam tree, a close relative to the American hornbeam and once more common in these parts. Since this small tree prefers damp defiles along the Colorado River, much of its favored habitat now lies beneath Lake Powell. Growing in a joint as it does, it represents those plateau province plants that thrive in cracks where runoff water is concentrated, and elongated roots can reach a long way to find it.

Other native plants take advantage of rain that sheets off the slickrock. Tucked into such a foot-of-the-cliff niche is another cliffrose. Its five-lobed leaves, dotted with tiny glands, are smaller than its creamy flowers with yellow stamens. Already some of the blooms are going to seed, hinting at the feathery plumes that, later on, make the cliff rose resemble Apache-plume and even clematis. The cliffrose served both the Fremont people and the Anasazi, who shredded its bark to make their clothing, mats, sandals, and rope.

While desert plants have evolved a great variety of ways to locate, conserve, and store water, we humans have not. A final, tepid swig from my last canteen tastes particularly awful. Now I'm running on empty, out of both food and water.

These three days have pointed to how our comfortable everyday lives separate us from nature. We often see splendid images of this canyon country but don't usually taste the grit. Videos on wild places tempt us to consume natural history while crunching popcorn beside natural gas fireplaces. Because they're so engaging and informative, such packaged surrogates may tempt us to accept the illusion that we live outside of nature. While the planetary environment deteriorates, such images can lead us to believe that all goes well "out there."

Blowing sand and hungry rodents have reminded me that the world wasn't designed for humans. Hiking during a sand-and-sleetstorm, I have felt the sharp bite of the wind, experienced the forces that sculpt this fantasyland. By baring their ancient bedrock, and especially by disorienting us and denying us water, the desert arouses primal emotions and rekindles self-awareness. Where there's no place to hide, we're forced to find ourselves. As I approach my van, muscles cramped from dehydration, my well-filed fingernails reveal experience not with fantasy but with soft-rock reality.

Interlude

The Old Friend that Smelled like Gin

Like a puffy cloud, a powder-gray tree nestles into the weathered gray slickrock. Nondescript. Yet this tree beckons, its gray berries dotting its dull-green foliage. As I approach, shoelaces catching on brush, I realize that it's "only" a common Utah juniper.

After that generic identification I focus on this individual juniper. Dead branches bristle from the otherwise rounded crown. Beneath the speckled form, strands of loose bark hang like frayed clotheslines. Behind these, gray and cinnamon-streaked bark peeks through the boughs.

I reach for contact. Grinding a needle between my fingers frees the scent of cedar. After rubbing a gray berry green, I whiff the aroma of gin. Then my tooth breaks the berry's surface to disclose its bitter medicinal taste. Gin, flavored by oil from these berries, is short for *geneivre*, the Old French word from the Latin *juniperus*, meaning "youth." These trees do stay green for a long time. Their wood, so often used for fenceposts, resists decay. Juniper seeds can lie dormant for years until a freeze cracks their tough coatings.

To feel younger, why not climb this old juniper? But when I grasp a limb, strands of bark break loose and drop me backwards! Brittle branches crack around me as I hit the sand, feet in the air, dust in my face, hands still clutching some of the scruffy bark. I look like a horse that's rolled in the dust. Pinyon jays jabber derisively nearby.

Wounded vanity put aside, I finger the bark in my hand. Its thick end looks like a rope that's been chopped off. Brown wormlike needles and dried berries litter the sun-mottled sand. Just above my face the browns of the bark look like creosote. Mustard-green lichen hangs from dead twigs. I feel cozy, like a boy hiding in the shrubbery. Soon enough, though, discomfort intrudes. As shriveled berries begin to feel like marbles, I sit up. Twigs gouge my scalp as stumps of bleached branches, each bowed with age, jab my bare back.

Why has this tree spurned my hug? The essayist Montaigne remarked that a defect in us "hinders communication between animals and humans." Surely, then, such limitations also keep us from relating to plants, our more distant relatives. With the five senses alone, intimacy eludes us, so intuition or imagination may offer better connections. Perhaps if we adopted the yoga posture of "the tree," standing on one leg with our foot rooted, we could feel more akin to trees. Or perhaps if we talked to them as we do to house plants, we might find relief from the estrangement that ails us, not them.

In *Desert Solitaire* Ed Abbey contends that he could write a book about not just junipers but about a single tree. Yet too often common plants become like robins, items we identify at a glance, generically, from a distance. We stop looking once we glimpse the robin's proverbial red breast, which is really not red at all. We carry our categories into nature, ignore the individual for the type, and then wonder why we're not seeing much that's new.

Cultural baggage dulls perception. By presuming to see this individual tree as an old acquaintance when I actually know only its species as a type, I might well have missed the individual. Contrary to what most of us learned in biology classes, to classify is not necessarily to understand. This is why Roger Tory Peterson, famed illustrator of bird guides, indicated that "I don't enjoy being out with birders—they want to identify a bird and get on with it. I like to spend an hour with one bird."

Naturalists can utilize both approaches. They can both marvel at the general characteristics of junipers and also use them to appreciate one twisted individual that's survived for centuries. What stories does its wood hold? How would this tree narrate the comic anecdote of the overly friendly homo sapiens? If we could connect with it, it would certainly enrich us. Imperceptibly, perhaps it already does.

3.

Enchanted Havasu

*The glories and the beauties of form, color, and sound
unite in the Grand Canyon—forms unrivaled even by the
mountains, colors that vie with sunsets, and sounds that
span the diapason from tempest to bubbling fountain.*
—Major John Wesley Powell

Hikers slump on their car bumpers, rubbing their feet. A
round-faced Havasupai Indian wants to rent us a pack horse,
but we'll carry our own packs. This is our third pilgrimage to
this exotic oasis, sculpted in baroque travertine and painted
with the dazzling colors of Grand Canyon.

As we zigzag down the bluff, the parking lot clatter
dissolves into the pristine air. No L.A. smog smudges the
canyon today. Ravens finger updrafts, their deep caws barely
audible. Yuccas with waxy flower stalks and cliffroses heavy

with blooms embellish the rocky slopes. As we stroll down-ward, athletes stride uphill, arms flailing. One red-faced fel-low presumes that we haven't hiked Havasu before: "You'll get there—it's worth it." Deborah reminds him that "We're here; no hurry!"

This first mile is indeed too spectacular to miss. All around are the familiar strata of Grand Canyon: the hard, light-gray Kaibab limestone; the gray, shaly Toroweap forma-tion; the buff-and-blond Coconino sandstone; and the red-dish-brown Supai sandstone. Below, more canyons cut into ancient layers dating back over a billion years. Fine vistas across Inner Gorge reveal the North Rim a dozen miles away. Grand Canyon humbles both temporally and spatially. Our lifetimes seem insignificant compared to its time spans, and we feel tiny compared to its enormous size.

Despite its isolation, Havasu Canyon attracted early Span-ish and Mormon explorers. When Francisco Garcés finally reached Havasu in the eighteenth century, he stayed six days with the Havasupai tribe. He left powerfully impressed with the Indians' "prison" and noted that interacting with the tribe "served to divert the melancholy that it caused me to see myself buried alive in that calaboose of cliffs." Spacious though it is, Grand Canyon has often inspired feelings of entrapment. Exploring the canyon a century later, Major John Wesley Powell hoped that after "a few more days like this we are out of this prison."

Still later Havasu interested the Mormons. To them, the natives were Lamanites, descendants of Laman, son of the prophet Lehi who led Israelites to the New World. Although the dark-skinned Lamanites bore a curse, the Mormons leaned toward conversion rather than extermination. They showed special interest in the Hopis who occupy mesas to the south. When Lee's Ferry crossing, the best way around Grand Canyon, hardly offered a direct route, they sought a shorter one between the Hopi villages and St. George, Utah. On one of the seven expeditions undertaken before 1872, Mormon leader Jacob Hamblin probed this canyon, but the Havasupai did not receive him warmly. Before releasing his party, the tribe demanded that Hamblin not tell other whites of their hiding place. In 1889 the Havasupai performed a Ghost Dance invoking the spirits of their ancestors to drive away whites. On occasion the tribe even fled their exotic home to hide in remote areas of the canyon.

Since Hamblin's time the Havasupai have remained free from major incursions, except for us hikers. Though white ways have changed them, the Supai still celebrate traditional feasts and dances for themselves, not for the tourists. Like the Hopi, they've retained ancestral lands largely because of their remote location. For centuries the tribe farmed its canyon oasis in the summer and gathered food on the plateau during the winter. Unlike their white visitors, their lifeways were geared to survival, not comfort. However, treaties diminished their hunting, gathering, and grazing rights, which disrupted traditional ways and left them dependent on the Bureau of Indian Affairs.

In the late 1980s uranium mining at Red Butte on the South Rim jeopardized both the people and their land. While one group of Havasupai filed suit, another trashed power lines. These monkey wrenchers pointed out that "a wealthy white elite" would profit from the uranium mining while Indians would absorb the birth defects and cancers. To the chagrin of cynics who think that environmental groups ignore minority cultural concerns, Earth First! became involved. Invoking the Havasupai name for Red Butte that means "the belly of the mother," thirty women danced in circles to reclaim this and other sacred places.

Novelist/activist Mary Sojourner writes that in the early 1990s tension rose with the controversial Environmental Impact Statement for the Canyon Mine. After interviews with members of the tribe, a Forest Service anthropologist somehow concluded that the Havasupai have "no discernible religion and no religious rights to the land." Actually the Havasupai had simply not divulged their secrets. In fact, each year the tribal traditionals meet near Red Butte to perform their ceremonies. With one culture guarding sacred lands and the other desiring nuclear power, the conflict pitted Native American against Euro-American values.

But today Deborah and I put politics aside. We descend into a canyon within a canyon, incised into the Supai layer by the stream that cut Hualapai Canyon. An Indian wearing a jean jacket and a black hat leads the tourists on horseback. Scruffy dogs trot behind. Prey in its mouth, a roadrunner trots across the trail and springs into a cholla cactus where it will later trickle the partially digested lizard into the mouths of its nestlings.

This five-mile section of trail channels between walls that become higher and steeper to expose seeps, streaks, and undercut cliffs. Wild cherry, Utah serviceberry, Mormon tea, and scrub oaks give way to feathery-green acacias, mulberries, and small cottonwoods, their lime-green leaves burgeoning from sticky buds. Western wallflowers emblazon the reddish sand, and orange desert mallows, still rolled in buds that resemble sweetheart roses, glow like miniature hollyhocks in the muted light. As our boots squish on wet sand, we're stepping into spring. Though many hikers view it mainly as a tunnel to paradise, Hualapai Canyon itself is a great hike.

Anywhere else it would be appreciated, but here it's over-shadowed by Havasu Canyon.

When we first glimpse the blue-green water of Havasu Creek, we break stride, astonished once again by its color. This arresting and enchanting stream, as David Lavender suggests, remains unforgettable:

> As a shield against the sunlight lancing off the red, there was green shade: massive-trunked cottonwoods, each heart-shaped leaf shimmering in the faintest breeze. And the water! The tourist catch-phrase for Havasupai, "Land of the Blue-Green Water," conveys the color, but not the luminous dip and glide, the whispering effervescence, or the compulsion to roll up one's sleeve, lie flat, and reach for the gleaming bottom of that stream.

Pink roots stripe the bright bottom where grasses waver in the current. Above the creek's shimmering "dip and glide," shiny cottonwood leaves flicker. Above them, cliffs soar skyward, red ocher against a deep blue sky. We lean into the sacred stream like pilgrims cleansing before they enter a shrine.

The creeks flowing from the South Rim toward the Colorado River are notably limey. The caverns just off Route 66 suggest how underground streams dissolve enormous quantities of limestone. This suspended lime causes the silt and sand to coagulate and sink rapidly to the chalky bottom, which turns the water an almost electric turquoise.

In this canyon, and on the Colorado Plateau generally, changes tend to occur either very slowly or very fast, either grain by grain or boulder by boulder in a flash flood. In placid times the carbonates settle out to become travertine limestone that forms scallop-shaped pools. As the creek bed changes course, curved shapes hang far above water level. Porous travertine absorbs pigments, especially iron oxide, that percolate down from overlying layers. Riddled with conduits, the tawny or rusty travertine drips and tongues enhance a surreal ambience, one implying that the travertine was once a part of the flow.

The home of the Havasupai is nestled in a verdant valley where clouds of painted lady butterflies feast on peach blossoms. Each year, especially in flowery springs, millions of these nomads head east, north, and west from the Sonoran Desert in Mexico. Their nomadic lifestyle makes them the most cosmopolitan of butterflies.

Far above Supai stand the Wigleeva, the sacred rocks. One is female, rounder and shorter; the other is male, columnar and taller. According to legend, these rocks determine the fate of the tribe: when they fall, the tribe falls. As one native woman expressed this belief, "Rocks all have their places out there"; they are not what we might call "inanimate objects." Nature displays balance and harmony, the Indians apparently believe, and humans must live in accord with it.

Lined with huge stumps, the path into Supai is deeply trodden. Cottonwood puffs fill the air and cling to rusty fences. A mule train trots by. Since the pace of life is slow, we stand out as we stride to reach the reservation office before it closes. As I loiter outside, the village seems both the same and different. Indians still lounge against fences, scrawny horses still slump beneath trees with chewed bark, and dogs still sleep in the street. This spring, though, a man plows with a tractor, a woman drives a Cushman cart, and a new Supai Lodge offers more amenities than the old cabins. The prices have changed, too: after we part with most of our cash for our permits, we have little left for burgers or tacos. But that's OK because our box, mailed weeks ago to avoid hauling food, should have arrived.

We saddle up, groaning under our packs, and trudge toward the campground. Where soft sand impedes tired footsteps, we wobble under our packs. Dust squirts from beneath our boots until we discover that the sand offers worse footing when we hike too fast. There's no curfew at the campground, so we'll get there when we arrive.

Despite our fatigue, Havasu Canyon works it magic. Large barrel cacti, which the Indians bake with agave hearts, stud the canyon walls. The creek glides through thickets of willow topped by box elders, ashes, and cottonwoods. Braided by froth, Navajo Falls cascades down a mossy cliff canopied with greenery.

Soon we hear the roar of a much larger falls: Havasu. We strain our eyes for a first glimpse of the arching falls and its enormous plunge pool. When Deborah sees them, she raises her hands to cover her mouth, which is speechless. Nothing is lost, since no words would do. We stare at its glorious color, an even deeper aqua than upstream. Tall prince's plumes luminesce like candles in the twilight.

After doing ten miles today we hobble into the campground. We finally unharness, rolling our shoulder blades, and yank off hot boots, ignoring laces, to massage our feet. Since it's almost dark, we spread our bags on the sand and drop off.

Breakfast is granola and coffee. Ah, caffeine! Life returns. Mug still steaming, I wander into the sagebrush to look for a bird that's been chirping. But instead of finding, I get found. High on a mount, the tribal policeman looks me over, nods, and trots off. It seems odd to have an officer follow you into the brush, so we set off for Havasu Falls. Perched on scallops of lime deposits, two Indian women contemplate the pools. I stop three times to compose photographs; Deborah waits, absorbing all the beauty, but she's irritated. And she's right. No more pictures. Photography antagonizes some members of the tribe and disturbs the sacredness of the place.

Above the falls is the reservation cemetery, its mounds covered with plastic flowers and skull-sized rocks. Nearby, a helicopter pad lies wisely abandoned. Nowadays the tourist choppers just bank and circle a couple of times—an improvement over the once-common landings though still an affront to the senses. Yesterday planes droned overhead much of the way. During peak tourist season, a thousand scenic flights a day clog the skies over Grand Canyon. Aircraft noise degrades the grandeur of the experience and denies hikers the sublimity of silence. Yet operators refuse to limit their flights or even to muffle their engines.

A faint trail leads toward a mesquite tree. Behind it, beside a notch in the canyon wall, we start our climb. When Deborah cannot reach a handhold, she stands on my shoulders while I brace myself, feet spread for stability. We check every hand and foothold before pulling ourselves up rocks polished by other hands and feet. Wildflowers sprout from the cracks in front of our faces. Hedgehog cacti sport deep scarlet blooms with yellow centers. Watchful for spines that could throw us off balance, we pick our way toward the rim where a cairn marks the bench trail.

From here the top of the Redwall limestone layer drops hundreds of feet to the creek far below. In contrast to the luxuriant canyon floor, this shelf is rocky, arid, open; one can see both the north and south rims in the distance. The cloud cover is blowing through, dappling the rocks and leaving mare's-tail cirrus swirls in a light-blue sky. Moved to eloquence, John Wesley Powell depicted such a vista as "awful in profound depths, sublime in massive and strange forms, and glorious in colors." In a paradox worthy of the *Tao Te Ching*, the canyon's enormities of space and time can occasion a dissolution of the self, yet the intensity of its sensations also energize the individual psyche. Exhilarated by the space, the light, plus the array of forms and colors, we can barely attend to our footing.

Up here the plant life is sparse yet vital. Adapted to this Sonoran Desert environment with bulbs that store moisture, wild hyacinths resembling wild onions bloom on the rocky slopes. Sotol, a kind of yucca, flourishes along with dwarf buckthorn, prickly pear, and barrel cactus. A grayish cactus wren sputters some staccato chatter, fidgets with a twig, and flits into a cholla cactus.

Small redbud trees add their own puffs of chartreuse to the drainages. These Western redbuds are rarities. In March, before their glossy, apple-shaped leaves pop out, these trees burst forth in light lavender flowers instead of the usual magenta. Unlike most legumes, redbuds bloom from both their trunks and branches, engendering small fruit that resemble snow peas. Some of last year's pods, now bleached and dried tan, still hang beside this year's withered blooms. The Diné, or Navajo, roasted these pods to make incense while other tribes used the twigs for making baskets.

As the trail skirts the cliff, Mooney Falls rumbles from its chasm. Below us lie sheer, smooth Redwall limestone cliffs, a plush carpet of greenery, and a thin ribbon of turquoise water. Through a jagged window in the rock, my eyes plumb first to the canyon floor hundreds of feet below, then another two hundred feet to the base of the falls. Hikers down there crawl like bugs, halting before the abyss. My muscles tighten. I ease back from this precipice, rock by rock.

In a drainage nearby, Deborah reclines on the smooth limestone, relishing the warm sun on her heart and the cool bedrock under her backbone. Nude woman, naked rock. We chuckle about the sign the Indians posted by Havasu Falls: "NO NUDE." Balanced in this yin and yang, her flesh sings an old Havasupai chant: "Make me always the same as I am now."

Back at the campsite we face a basic reality: if we want to eat tonight, we'll need to hike the two miles back to retrieve that box. So we put new moleskins on our blisters and set out for Supai. Vining snapdragons that resemble wild morning glories drape the canyon wall near Havasu Falls. Nearby, waxy prince's plumes bend with the gusts, but this doesn't discourage a rufous hummingbird from hovering in perfect unison with their sways.

Alongside desert willows with orchid-like blooms, ravens flick leaves. Croaking deeply, they take flight only as we encroach within fifty feet or so. At first they flap their wings like a crow, then they soar like a large hawk. Near Supai, where bright green fields contrast against red canyon walls, scrawny mares stand under a spreading tree, their swayed backs worn hairless. The Indians once raced their horses along this trail.

Much to our relief, we find the post office open. Praise the tribal gods! But what if our food has somehow not arrived? The mail comes by mule train, and our box would hardly be priority mail. We're almost out of money, and this isn't a place that runs on plastic, either. Luckily, when the clerk can't find our package, Deborah spots it. Now we've got money to spend. Moments later we're tearing into an Indian taco made with thick fry bread.

The trill of the canyon wren and the twitter of a yellow-throated Western kingbird accompany our return to camp. As we gather white-blossomed watercress, frogs bulge their throats and dragonflies bank and veer. Wild celery, a perfect miniature of the cultivated variety, flourishes under a willow. Cress is an Old World plant that's spread to even the most remote desert canyons; onion, garlic, and celery are among the many New World plants that enhance the Anglo-American diet. Other edible native plants, however, remain largely unknown to non-Indians: salted, roasted pinyon pine nuts are as tasty as potato chips and certainly more nutritional.

Distinct rock layers gleam in the waning daylight. A huge century plant looks small against pink clouds in a baby-blue sky. Violet moments flash and fade. Finally the sky blazes a brilliant orange that commingles into a grayish blue. Neglected, our fresh watercress and celery soup almost boils over.

Dawn arrives in a burst of birdsong. When I can't see any birds, I settle for a bug. Just outside our tent, black *pinacate* (Spanish for "the presumptuous one") beetles are thrusting their rears into the air. Several dozen species of these stink beetles thrive in the Southwest. Rather than run when disturbed, their noxious-musk defense has earned them the local designation of *"pedodos,"* or "farters." Their black color seems poorly adapted to desert habitats, but pinacates burrow during the heat of the day. Their waxy cuticle also reduces dehydration, and a dead air space beneath their wing covers insulates them from extremes of temperature. But these beetles' spray couldn't match my boots. No need to check for scorpions, tarantulas, black widows, or brown recluse spiders this morning, since none of them could survive the stench.

Today our first stop is Mooney Falls, named after the miner who fell to his death here in 1880 while trying to get down this one hundred-and-ninety-foot drop. "The Mother of Waters," the Indians' name for this magical spot, figures strongly in Supai beliefs. So rooted are they to their canyon that they believe the spirits of the dead congregate here, rising and falling with the mist. With this in mind, it's even more outrageous to find that in the 1920s government and industry colluded in an attempt to generate power here. Fortunately a flood washed out the machinery before the power plant could desecrate this sacred place.

Mooney Falls still stops many hikers. While some don't know about the trail that miners blasted out of the travertine, others fear the heights or the shaky chains that serve as guard rails. The path switches back and forth, over and under limestone bowls where it enters a tunnel and emerges onto a landing. From here, looking down is quite literally breathtaking. My chest constricts as my whitened knuckles lock onto the cold pipe. Each step becomes deliberate. Trodden limestone is slippery even when it's dry. This descent ends among giant scalloped tongues formed when the creek followed different channels. But these tongues were created only yesterday compared to the nearby Redwall limestone, formed 350 million years ago.

Since Mooney Falls faces north, we emerge into the deep shade of its chasm. Our eyes track globules of water that whorl downward, then disintegrate into beads from the blue. With its crystal pools this basin is both vast and intimate: after gaping at a thousand-foot rock wall, we gaze down on a tiny toad.

A "Japanese garden" in a box canyon ascends in pools trimmed with orange monkey flowers and lacy maidenhair ferns. In this exquisite grotto, frothy waters slip between plants and spill over sandstone lips. Water striders ripple the pools. Below one spillway a dark-gray water ouzel (dipper) dives, swims with its wings, and pops up. As it dips itself dry, its dusky plumage absorbs the sunlight. A nest shaped like a basket turned on its side hangs nearby. As the parent arrives with grub in mouth, the nest explodes with cries. Four bright yellow beaks, each flared wide open, instantly fill the ring.

Below Mooney Falls the environment becomes more pristine because the hikers thin out and the Indians can't bring horses down this far. The creek cascades into pools, each a living spectrum of pastel watercolors—the aqua of the creek, the burnt oranges of the cliffs, and the bright greens of the sunlit trees. Snakeweed bristles from the banks to link itself with the fossilized horsetails that grommet the red limestone. As living evolutionary success stories, some plants have changed little in three hundred million years. Cozy in this leafy glade, we inhale enriched oxygen from the foliage and exhale carbon dioxide for the plants to breathe. Our lives partake of life's reciprocity.

As it streams over a lip, the crystalline current purls into foamy pools. Baroque travertine lips sometimes curve completely around, actually spilling upstream. Pulses in the current cause one mini-falls, then another, to murmur and wane like waves marking the rhythm of the flow. With its quiet pools, Havasu Creek shimmers with painterly images of the cracks reflected from the canyon wall: cracked water, wavy rock. If he had done his light and shade tableaux here, Claude Monet might never have returned to Giverny.

The canyon's walls rise higher. Its floor broadens as the trail snakes through knee-high grasses and wild canyon grape-vines. Fewer cottonwoods line the creek on this stretch because beavers are so proficient. It takes them only a few nights to gnaw around a trunk well over a foot in diameter, yet they often chew through the cambium layer and leave large trees to die. As cottonwood numbers dwindle, will the beavers kill more large trees to feed on their inner bark; that is, will they eventually consume their primary food source? Responding to this prospect of instability, Deborah recounts how a Diné woman once told her of a tribal belief that chaos is the natural state, that we become unhappy when we try to impose order.

Though Deborah says she's tired, I urge her to keep going. Nearly three splendid miles below Mooney, the canyon narrows for Beaver Falls. From a ledge we marvel at a spot where the stream takes three sharp turns to create one of the deepest, bluest pools. It's time for full immersion in Havasu Creek. As I slip in, a trout flashes under a ledge. Underwater my eyes taste the blue-and-bubble world of the plunge basin.

Moments later a wild man parts the foliage. As W. L. Rusho describes Everett Ruess, the "vagabond" of the 1930s, this fellow "could almost resonate with the light waves that struck him from all points in the landscape." In his tattered cut-offs, Robin is muscular, sunburnt, and bearded, more vagabond than aesthete, more hearty than savage. The wrinkles of the landscape are etched on his face. His bleached-and-matted hair makes us look like blow-dried tourists who have just stepped from a helicopter. This canyoneer has lived here for a couple of months, he says, scrounging meals where he can. For five weeks he's subsisted under a cliff. He catches a few fish and clubs a few mice but mostly eats what the Indians traditionally ate—agave hearts, prickly pears, and wild grapes. Minimal-impact living.

Committed to personal liberty as he is, Robin suspects governments and their agents. "Seen that tribal cop?" he asks.

"Don't know if he really wants to keep drugs off the 'res' or if he sells the stuff he confiscates." The hallucinogen peyote has long been a subject of debate among the Havasupai. Was it the cause of unweeded gardens or were the dreams it evoked truly sacred? An earthy man at home in the earth, Robin bounds over rocks in sneakers with flapping soles. When living off the land gets old, though, I'll bet he grabs an Indian taco at the Supai Cafe on his way out.

On our return up-canyon, an orange-banded kingfisher squawks at a pied-billed grebe that competes for the same small fish. The grebe dives under as the kingfisher becomes a blur of blue and white. This grebe bobs up shaking its head as the kingfisher jaws loudly. Such harsh staccato sounds, or "rattle calls," are the kingfishers's way to scream "get out of my territory!" Along water courses around the globe, these bold birds habitually stake out a stretch and confront any intruders.

On the bank a white-haired fellow is digging up plants and stuffing them in a bag. He immediately informs us that he's a botanist gathering specimens of rare liverworts. When I ask whether it's a good idea to collect plants already sliding toward extinction, he bristles: "This is my specialty, you know." Is the traditional, scientific practice of bagging rare specimens for study still tenable today, I wonder. Collection surely contributes to extinctions, leaving pickled animals and dried plants as unsatisfying laboratory curiosities. Showy lady's slippers have already been picked and dug to extinction in Acadia National Park, as well as in other places. The last passenger pigeon died in a zoo.

We sit on the stream bank, feet in the bubbling water, savoring the canyon wall. Their wings back-lit a gauzy white by the sun, doves flock on the rim far above. Encrusted salts tint the red rock with pastel oranges and pinks. A band of ferns trims the waterline. Below this fern curtain these vibrant colors trickle into the shady creek where, with the whip of its tail, a fish roils the reflections from the white bark of a palo blanco tree.

Supper brings talk about self-sufficiency. Robin's hunter-gatherer ways highlight the paradoxes of backpacking. We haul civilization's amenities on our backs, nicely miniaturized. We go back and forth between civilization and wilderness, making brief forays dependent on what we can carry. Like Robin we may eat native plants, picking only the abundant ones, but we also rely on that crumpled box from home. Robin has greater freedom and a closer relationship to the land, but we have less impact. One way or another, everybody leaves footprints in the wilderness.

Tonight, under a clear, moonless sky undimmed by pollution, the densely packed stars shine many-pointed. Even on a clear night, though, the several hundred thousand visible stars constitute but a tiny percentage of the billions thought to comprise the cosmos. In fact, many of the "stars" we see are whole galaxies. Brilliant English physicist and cosmologist Stephen Hawking estimates that "our galaxy is only one among some hundred thousand million that can be seen using modern telescopes, each galaxy itself containing some hundred thousand million stars." When we ponder the magnitude of the cosmos, how much significance can we attach to our egos? There's nothing like a starry night in Grand Canyon to humble the beholder.

At sunrise a Western wood peewee sounds its wake-up call. Hoping to attract a mate, males sing their nasal dawn call every morning during the nesting season. Its "pee-wee" triggers recollections of the tree house I once built where the peewees woke me gently, unlike the cardinals that whistled right in my ear.

With no one stirring in the campground, Havasu Falls is still deserted. The air cools with each step down the trail toward the plunge pool. Havasu's round basin is ringed by huge cottonwoods, their trunks sprayed the same fawn color as the sand. The creek plunges down a chute surrounded by baroque tongues of travertine. Below them, behind the falls, constant spray nourishes lush maidenhair ferns.

Braving the chill I creep into a cave where my breath clouds. Waters gurgle through the travertine, gradually filling up its conduits. This must be what life sounds like to an earthworm or a mole. The underground trickles bring me under their mythic spell. Thoughts of death seem sexually exciting; I imagine life in a womb, even death in a tomb. Then, spooked, I bolt back into the daylight, short of breath, huddled away from the spray. The falls are even more dazzling to eyes opened wide in a cave.

Teenage boys thunder down the trail, but soon the falls spellbinds them into silence. They sit down, push back their caps, and stare. Arguably the most picturesque place in Grand Canyon, Havasu Falls is a visual paradox. From the trail skirting its west side, it seems to be an enclosed amphitheater, but from below it seems quite open. On the east side, across the creek, Carbonate Canyon twists for nearly a mile toward massive bowls glazed by falling water and rock.

Departing is difficult. Before we set out for Supai, we fill our canteens and bellies with bubbling water from Fern Spring. A campground mutt follows us to the village, lunging with real fierceness at lizards. When our canine companion sees a foot-long chuckwalla lizard, he charges. The slow-moving chuckwalla, however, proves faster than it looks; it dives under a rock. Rushing headlong, the mongrel skids to a halt like Goofy to avoid a big barrel cactus.

Later, while we eat lunch, a gray fox trots by on the opposite ledge, nose to the ground, never giving us a look. It's thrilling to see a fox up close, especially where the semi-wild dogs could give chase. Such pursuit could prove fatal since foxes aren't fast—they stalk their prey and pounce like cats. Still later we linger at a wild cherry in full bloom, then touch the lacy new leaflets on an acacia, bright green against the pocked sandstone. As we surface from the trench cut by Hualapai Canyon, a sage thrasher puffs its brown-speckled breast, cocks its head to brandish its curved beak, and bolts, flashing a whitish tail.

The last mile is tough, especially when we reach the switchbacks in the afternoon while the sun hammers hardest. The great walls of stone sway in the heat. To forget the pain I play a foot-and-knee game, zigzagging from one side of the trail to the other to lessen the incline. At the same time I take small steps and push my knees back to ascend by straightening my legs. My lungs are syncopated with my legs. I inhale while taking four steps to the right, then exhale while zagging four steps to the left. This diverts attention from my pack, which chafes my shoulders and hips. Even hiking sometimes rewards concentration on technique.

Heads down, ears full of mechanical drumbeats, joggers nearly run us off the trail. It's hard to imagine coming down here without free eyes and open ears. In a place like Grand Canyon, this borders on sacrilege. Granted, running as a way to reach otherwise inaccessible places could be justified. But using nature as an aerobic workout arena puzzles me. These sweating, panting fitness freaks race right by it all, out of touch with what lies all around and even underfoot. Most aren't here to marvel at the colors or to hear the birds. More often they're going for their personal best—a telling phrase that may suggest both self-absorption and a competitive orientation. Compulsive conditioning defines the body as a machine, a material thing commanded by, yet estranged from, the mind. After centuries of debasement, it's time to respect our physical being without joining the cult of the body.

Just below the rim, Coconino sandstone enhances the quality of the light. This tawny layer—possibly the loveliest in the Southwest—crops out all along the eastern end of Grand Canyon. Composed of nearly pure white quartz grains, the blond Coconino stains easily: its sheer, smooth cliffs are dripped with coffee, chocolate, and cinnamon. It is rippled by winds that blew 270 million years ago, illustrating how changes in wind direction enabled dunes to accumulate, one upon another. Beside my pack a fossilized footprint provides a connection to a Paleozoic animal that walked here millions of years ago.

As I approach Hualpai Hilltop, occasional trash suddenly jars the senses. Reentry symptoms are a good sign, however. If you don't feel them, you probably haven't been away long enough. Soon we're slumped on the floor of our van, still under the spell, happily glazed as other hikers lumber by.

Far below us Havasu Canyon throbs with life. Its streams pulse through rock like blood courses through flesh, its creek is always depositing lime. Flash floods periodically sweep it away, even the travertine scallops below Havasu Falls, and the cycle begins again. The blue-green water seeps, trickles, swirls, cascades, and thunders, enchanting thousands of pilgrims a year.

Interlude

Sacred Places on a Living Planet

Some places have long been thought to reverberate with special, possibly divine energy. At Stonehenge, Delphi, Palenque, and Machu Picchu, people apparently settled nearby to experience the benefits. Other spiritual centers have been left undeveloped. These include Mt. Fuji, the Black Hills, Mt. Rainier, Mt. Shasta, and Yosemite Valley—plus Zion Canyon, Grand Canyon, and the nearby San Francisco Peaks, sacred to both the Navajo and the Hopi.

The holistic spirit of most American Indian religions, which regards all places as sacred, often assigns special significance to locations associated with creation events or ancestors' spirits. Southwest tribes do not simply react to perceptions of spiritual energy. Instead, their medicine men often designate "places of power" based on tribal stories.

The Navajo, or Diné, as they prefer, account for human creation in a wonderfully earthy way. They tell of First Man and First Woman who, traveling with animals and insects, surfaced from a world of spirits within the earth with the help of a friendly badger. Many Diné locate the place of emergence into the Fourth World in northwest New Mexico, where First Man and First Woman found a baby on Gobernador Peak. Here in the center of *Dinétah*, the holy birthplace of the Diné, this baby miraculously matured in just twelve days to become Changing Woman, who then mated with the sun to conceive twins at nearby Huerfano Peak. These sites remain sacred to the Diné.

In the Hopi world view, on the other hand, links with

creation and the energy of a place are both important. Like the Diné, the Hopi revere their *sipapu* or place of emergence from the earth. In Hopi mythology the earth's guardian is a wise Spider Grandmother aided in her duties by two young men, one at each of the earth's poles. At the North Pole sits Poqanghoyam, working his magic to maintain order. At the South Pole sits Palongawhoya, beating rhythms on a sacred drum. These rhythms, many Hopis believe, channel vital energy to bring the planet to life. The Hopi dances, in which drums thunder while feet beat the earth, express this rhythmic connection. For the Hopi the earth's energy surfaces most strongly at certain spiritual centers recognized by the elders as sacred places essential to harmony and healing.

Many cultures have sensed that energy radiates, or that an aura is exuded, at certain locales, but until recently little scientific research has probed the conditions associated with these places. One study discovered that electromagnetic currents exist around the San Francisco Peaks, where the Diné and Hopi believe their kachinas reside for most of the year. These currents produce an abundance of negative ions and are said to energize the body and soothe the mind. At other sacred sites, natural uranium is present, leading one to wonder whether it is radioactive decay that enlivens the air. Waterfalls, which have long been thought to exert a spiritual presence, also produce such ions.

While rainbows add enchantment to waterfalls, rainbows in stone—arches or natural bridges—also convey their own mystique. Rainbow Bridge astride Navajo Mountain on the Utah/Arizona border holds special meaning to the traditional Diné creation stories. According to Clyde Kluckhohn's *Beyond the Rainbow*, not only did their Rain God reside here, but their Sky Father created the Bridge to rescue a "hero god" from a flash flood. Later he transformed the rainbow to stone as an emblem of omnipotence and watchfulness over his earth children. Since this place of the Rainbow Spirit must be approached with reverence, Diné tradition asks visitors to offer a prayer before stepping into the shadow of the Bridge.

Sometimes, however, the significance of a place may imply more about culturally imposed perceptual limits than about the possibility that some places are more holy than others. When our culture designates sites as sacred, they're usually tied less to natural characteristics than to human history, such as the hallowed Judeo-Christian shrines or the battlefield at Gettysburg. If we see only our own cultural-specific sites as sacred, we reinforce the compartmentalized thinking that allows exploitation of what we consider non-sacred parts of the planet.

Some locales in nature, however, do exude a special magic. As a place where the rock itself seems to pulse, where water courses through the veins in the rock, Havasu Canyon on the South Rim of Grand Canyon exemplifies the Gaia hypothesis. Named for the long-neglected Greek goddess of the earth, the Gaia theory sees the earth as a living (though non-sentient) entity. This is a radical idea, for Euro-American culture sees the earth as basically lifeless air, water, and rock with only a veneer of life. (Two notable exceptions were the beliefs of New England Transcendentalists and early Mormons, both of whom perceived a spiritual dimension to physical reality.)

Gaia thinking also emphasizes the divine feminine that once animated early Greek, Minoan, and Mesopotamian cultures. Ecofeminist studies have shown that both Middle Eastern and Western civilizations underwent fundamental shifts from female, earth-oriented deities to male, sky-oriented gods exemplified by the Greek Zeus and Hebrew Yahweh, or Jehovah. In one graphic case, Uranus, the upstart Father Sky deity, raped Gaia to mark this ascent of male power. Classic Greek culture turned away from the earthy and female when, in a search for universal essences, Socrates and Plato downgraded the physical world.

Thus, as so many before me have contended, the Platonic tradition contributes to our ecological crisis. The notion that the physical world is a lesser reality causes many problems—bad metaphysics may do us in. The redoubtable Edward Abbey rebutted Plato well: "What ideal, immutable Platonic cloud can equal the beauty and perfection of any everyday cloud floating over, say, Tuba City, Arizona, on a hot day in June?"

Having absorbed a great deal of Platonism, the early Church Fathers condemned the residual Greek vitalism, the belief that spiritual energies inhabit everything. The early Christians also, as D. H. Lawrence pointed out in "Pan in America," transformed the Greek god whose body was part goat and part man into their devil, complete with the traditional black face, cloven hoofs, horns, and tail. By association, anything animal or sexual became bestial and evil. Viewed as the literal embodiment of the profane, the earth could hardly seem sacred. Once again this negative association contrasts with Native American stories, including those of the Paiutes who believe that Coyote gave birth to man and woman and taught woman how to give birth.

As advanced by British biologist James Lovelock, Gaia thinking seems to contradict the scientific assumption that the planet is comprised mainly of "inanimate" magma, rock, water, and air. But the Gaia hypothesis gains acceptance among scientists who are willing to define life differently. If

life involves self-regulation, then the earth itself is alive because, ever since green plants increased oxygen levels, its
systems have regulated atmospheric gases. If oxygen levels
exceeded the present 21 percent, fires would rage out of
control; if they fell much below that, many organisms could
not carry on their normal processes.

Gaia thinking encounters problems, however. It tends to
perpetuate both the stereotype of the "lowly" Earth Mother
and also of its counterpart, the "lofty" Great Father in the sky.
In addition to reemphasizing the physical, reproductive aspect
of the feminine, critics note that Gaia also deludes us into
assuming that surely Mother Earth, no matter how abused,
will always feed and clothe us, carry away our wastes, and
regulate global warming.

In addition, when we conceive of the earth as an organ-
ism whose systems work toward stasis, other questions arise.
How do we know when we are pushing the limit? If our
impact exceeds the earth's abilities to self-regulate, can we
confidently turn to ourselves, as the "brains" of the planet, to
restore equilibrium? Finally, the systems approach of Gaia
thinking may represent still another conception of the world
as a machine—a view that inhibits what intellectual historian
Martin Berman calls "the re-enchantment of the world."

But the best Gaia conceptions do merit consideration.
Few ways of understanding our relatedness to the earth sur-
mount the "Gaia meditations" of John Seed and Joanna
Macy: "Inner oceans tugged by the moon, tides within and
without. Streaming fluids floating our cells, washing and
nourishing though endless riverways of gut and vein and
capillary." These and other Gaia metaphors appeal to many of
us because they reduce our sense of isolation, as living beings,
from a dead planet and cosmos. And beyond metaphors, the

mystique of sacred places continues because many of us definitely do resonate with the energy of earth, water, and sky.

4.

Walden West: A Cowboy Cabin

*What sweet and tender, the most innocent and divinely
encouraging society there is in every natural object . . .
There can be no really black melancholy to him who lives
in the midst of nature, and has still his senses.*

—Henry David Thoreau, *Journal,* July 14, 1845,
just after arriving at Walden Pond

At its head, nearly fifty meandering miles from its mouth,
Dark Canyon cuts into the west shoulder of the Abajo (or
Blue) Mountains that rise from the desert of southeast Utah.
Described by Mormon pioneer Platte D. Lyman as "rough
and worthless," Dark Canyon was one of the last areas
claimed for grazing in the late nineteenth century. As Rod
Greeno contends in *Wilderness at the Edge*, it remains "argu-
ably the wildest canyon in southern Utah."

Barbara and I stride down Kigalia Canyon, packs squeaking in cadence. Soon we enter Peavine Canyon where bare, sculpted sandstones line the sunny side and well-spaced conifers creep up the shady cliff. Their long, slender needles glistening, giant ponderosa pines border the sage flats. As the canyon walls rise, their bands of beige and rust parallel the orange-and-black trunks that in turn pick up the tans of the long needles underfoot. A black and beige flicker swoops down to an anthill, feathers flashing salmon pink. Everything intermingles.

Thanks to old roads, this is easy hiking. Elk Ridge, really a plateau, became accessible when early settlers established a rough track from Blanding on the east to Bears Ears, two distinctive mounds, on the west side. A rugged road later crossed the narrow Notch and extended down the plateau rim to Natural Bridges National Monument. We pass an old corral, its boards warped and bleached by years of weather. While yesterday's ranching seems quaint, today's grazing degrades the wilderness. Hooves pit the mud along the creek, and the critters themselves become impossible to ignore. Their splats, bellows, and thundering hooves contrast with the pristine trees, cliffs, and skies. Rather than waiting for us to pass, these dogies low and thunder along just ahead of us. Barbara sputters about "sacred cows at the public trough." I ruminate on the issue of breeding the wildness out of animals, only to lose touch with our own.

At the confluence of Peavine and Dark canyons, we locate a spring. Below here, despite the old road, Dark Canyon becomes more remote and pristine. Peregrine falcons dive from cliffs, crystalline waters drop into deep pools, and Anasazi ruins see few visitors. Before long noisy pinyon jays descend to see what we're eating. As year-round residents, these jays survive the winter and even breed before the snows are gone. Like nutcrackers, they can remember most of the places they've hidden nuts in the fall. In addition, like most other Western jays, they maraud the nests of other birds to feast on their eggs.

The mooing subsides, but the cattle are loitering, heads turned toward us in a backcountry stare down. If we're not going to eat their dust, we've got to slip around them. To "head the herd," as the cowpunchers say, I'll need to sneak out of sight—otherwise I'll get outsmarted. Like a buckaroo in an old movie, I pull down my hat, scramble up the rocky slope, and pick my way to avoid either spooking them or gouging my flesh. Then, as the bovine beasts swing their tails, I charge down the slope, dislodging rocks, waving my hat, and mustering my best rodeo holler: "Heeee-Hawwhh!"

Unimpressed, the cows just chew their cud. Then I quit the "Git-along-little-doggies" imitations and just bellow from down deep. This time the herd turns tail and lopes off in a cloud of dust. Not only have I tapped into my primal energy, I feel as though I've done something mythically American. Western movies, cowboy comics, theme parks, and rodeos all perpetuate a complex mythology that brands many minds. Many of us love cowboy lore but don't love cows.

To my suprise, upper Dark Canyon differs from Peavine. Its larger stream cuts a salmon slash through blue-gray flats where scarlet paintbrush flicker inside the blue-gray sage. The rutted road leads upcanyon toward a broad chute. Here the stream splits into mini-cascades where plunge pools and sculpted ledges offer ideal cooling and sunning.

As Barbara huddles behind a curtain of falling water, I vault up to the ledge above and dam the flow with my body. She yells "Hey, you!" in mock protest. As the sunlight penetrates her cool hiding place, I lift my torso to release a tidal whoosh. No longer in a mood to frolic, she gives me a "grow up" look. Really, though, isn't nature a fine place to let the child out to play?

As we munch apples, skies darken overhead and thunder rumbles across the mesas. Before long the stream flows brick red, crested with flotsam and scum. A pink mist wafts from the pools where, only moments ago, we played in the sun. Big raindrops splatter in the sand, inscribing dark eyes with lashes. Despite the deluge, red dust shoots from under Barbara's boots. Moments later, shafts of rain link earth with sky. Screened by the stringy curtain hung by the downpour, the canyon walls look ghostly, slick, and silvery. Lightning strikes a ridge. While the stark flash lasts, my eyes scan the cliffs for waterspouts. When we skid into a gully, hands dug into the wet sand, the torrent is running higher and redder. Sticks snag around our ankles, making it difficult to plod on.

As the raindrops explode into spray, graying the sage flats, we can't see far enough to find an overhang. Hail stings my hands and pounds my raincoat like snapping fingernails. What pain animals must endure during hailstorms. Plastered to our legs, our jeans are smeared with red mud. This could be a cold, soggy trip.

About a mile upstream, beyond towering sandstone monoliths left by ancient meanders, we glimpse more melon-colored spray. A spillway floods another sandstone ledge, then thunders into a swollen pool about thirty feet below. Cold and lonely, I stare into the turbulence. To reduce this isolation I imagine the rain drenching millions of excited root hairs across the desert. As the moisture reaches the roots, every cell rejoices.

Finally the downpour lets up. Through the mist we see a corrugated roof that shines like a mirage. Someplace dry.

Suddenly our boots churn like bulldozer treads, struggling up the sandbank. Since the door's unlocked, we walk right in. Our boots clunk across sole-sanded floorboards, leaving blobs of red mud. Barbara opens a cupboard stocked with moldy flour. Mice have jumped several feet up to gnaw on anything available; apparently they scurried in here only to face winter starvation.

In no time I'm immersed in the past. A yellowed program from a horse show languishes in the dust and mouse droppings. An ancient cast iron stove with polished plates that read "Wrought Iron Range, Home Comfort, Patented" dominates the cabin. Alongside sits a freight box of firewood. Our everyday high-tech lifestyles—not to mention our wet clothing—ready us to delight in something simple like making a fire. Flames soon lick the burner holes while juniper smoke curls through the cracks in the iron. As the stove crackles and roars, my ears resonate with the pitter-patter on the rusty roof. Basking in a bygone era sure feels cozy, even if it's historically misleading.

While the Wild West promised settlers freedom, plus an opportunity to reinvent themselves, isolation and natural obstacles often forged new shackles. Only since the 1970s have historians seriously documented the disappointments—the broken axles, broken bodies, broken hearts. Was it because most settlers believed religiously in Manifest Destiny, in a national calling to spread civilization, that they could endure the hardships? Or were these tough-as-rawhide descendents of the puritan just much tougher than we are?

Today we forget that ranching pitted humans against bears, wolves, and mountain lions that prowled these canyons and killed not only livestock but cowboys' horses. Out of anger and fear the farmers, ranchers, and miners hunted or trapped these predators for bounties. These guys were not only tough; they were also mean. Their almost evangelical conviction against "varmints" reveals a hostility to other species that nature lovers deplore. But unlike urban environmentalists, these ranchers lived close to the land.

In 1879, while the Mormons' Hole-in-the-Rock expedition struggled to reach Bluff, south of here, Spud Hudson brought the first two thousand cattle into these mountains. Stories of big profits led to the establishment of ranch kings and cattle barons. Among the Blue Mountain grazers were the Carlisles, an allegedly aristocratic British family that bought out Hudson and forced smaller operators to run their stock in even more remote locations. In a move that foreshadowed the history of Western ranching, the Carlisles enclosed a large area of public land in Wyoming. By the mid-1880s the Scottish Carlisle Cattle Company, now one of the largest outfits in the country, drove nearly eleven thousand head to market.

Since the British fronted the capital needed to build Western railroads, British cattle barons soon dominated vast areas of the American West. This led to the establishment of British stock such as the Hereford that are poorly adapted to arid terrain. By 1887 Texans began to encroach. One observer described a "great brawling herd a mile long straggling down the river through Bluff" on their way up here to Elk Ridge. Before long local cowboys were roping "Texas renegades" that got lost in the canyons. In winter hired hands from different operations mounted drives to rid this enclave of wild longhorns.

According to the Forest Service pamphlet in my hand, this is Scorup Cabin. John Scorup settled here in the 1880s, summering his huge herds here. A card tacked up nearby reads: "Left Dark Canyon Camp this morning: headed for sweet ass hills. Be back sometime next spring. —John Scorup Oct. 18, 1904." The flip side reads: "Clyde Argyle, Utah Dept. of Agriculture, Car Radio 35 ALPHA 2." It's a good idea to turn over a government man's card.

When the Forest Service established the Monticello Reserve in 1907, it did not include Dark and Woodenshoe canyons. As a result, Dark Canyon became an open back door to the Elk Ridge plateau, the prime summer range where the Forest Service was attempting to collect minimal fees. Evasions aggravated the overgrazing problem here and reflected the ongoing uncooperative posture of ranchers.

While Barbara toasts her socks on sticks, I peel apart the damp pages of *Walden*: "In the midst of a gentle rain . . . I was suddenly sensible of such sweet and beneficent society in Nature, in the very pattering of the drops, and in every sound and sight around my house, an infinite and unaccountable friendliness." Thoreau's storm must have been less fierce than ours. Still, gazing out at the glistening scene, I resonate with his feelings. My face presses against the dirty window where my breath compasses a circle on the cold glass. As I gaze at the soft-focus scene, I hear Henry David whisper his enticements to the wilds outside. It's time, he urges, "to suck all the marrow out of life." For guys like us, that means getting outdoors.

The delicious aromas of moist sage and juniper greet me at the door. The hail has bruised the foliage, releasing their herbal essences. Our noses, so often clogged by soot and bludgeoned with industrial pollutants, can reopen to full sensation. Tendrils of vapor are curling like ghosts up the canyon walls, much as they did on the hills above Walden Pond. At my feet the deer prints are so fresh that their centers are still dry.

Once the fog has risen, I mosey around the spread. At the foot of streaked slickrock cliffs, glistening ponderosas tower a hundred feet high in spacious groves, their shade neither too dense nor too thin. Tucked beneath their long branches, box elders add puffs of bright green. In the foreground a firecracker penstemon waves its scarlet bugles against the greenery.

Suddenly a hummingbird buzzes—its ruby and emerald throat looking sequined—and stabs its beak into the penstemon's bugles. Beating about seventy-five times a minute, its wings blur almost invisibly. Hummingbirds maintain this high-energy output with the largest heart relative to body size of any animal on earth. With a whir the ruby-throat shoots right up to my face, brazenly looking me over, before darting toward a bare juniper branch. Despite their need to visit hundreds of flowers each day to maintain their metabolism, hummers seldom pass up a chance to indulge their curiosity.

This tiny bird now roosts in a dead tree that also hums with life. Lichens cling to its skeletal boughs, ants burrow into its trunk, and a hard-to-see horned toad is feasting on the ants. Dead trees house life; they're not just waste to remove for firewood.

In the palpable twilight everything is lush, soft, pastel. A haze lingers just above the sage flats to render them still more gray. Like Thoreau arriving at the pond, I feel a "slight insanity" of mood, a sense that, yes, "every little pine needle expanded and swelled with sympathy." Large droplets bead hairy leaves. The creek once more runs clear. Along its banks tall grasses are waterswept, flattened like the quills of a porcupine. Irish air and dreamy light are bathing arid southern Utah. This whole canyon, miles of it untrammeled and unspoiled, becomes ours for an evening. We experience a sense of freedom that was probably not as available to our more utilitarian predecessors.

Where a spring spouts from a grassy bank, crystalline waters soon tinkle into my canteen, the pitch changing as it fills. Reaching for some watercress, I come eye to yellow eye with a shiny frog that imbibes the moisture through every pore. I twitch but the frog stays put. We stare at each other for some time, the frog's grainy throat quivering steadily while my pulse calms down.

Even breathing disturbs the silence. One can hear so much from so far. Swallows—or are they mountain blue-birds?—twitter way across the sea of big sage. A chorus of distant songbirds continues the concert until dusk. This is a complex assemblage of singers, a more nearly intact ecosystem. How is it that any combination of birds, unlike just any combination of human voices, always sounds harmonious? If Henry David knew, he didn't let on.

Soon the starry sky arches like a floating vault, serene and seamless, so near and yet so far. In the distance an owl hoots softly, its sound almost palpable, as our eyes close and we simply listen to the night. This scent-drenched air evokes Thoreau's moonlight walks, which are rich with fragrances. Of all the senses, smell triggers the most precise associations. For me, the scent of linden trees in bloom always recalls a tunnel of trees on a country lane in Tuscany. Unfortunately, though, most of us can identify only a few dozen among the thousands of the scents our noses bring in.

At dawn a red-shafted flicker drums on the cabin. Since flickers defend territory about a half-mile square, drumming tells any intruders that they're entering a territory. These woodpeckers hybridize easily: the red- and yellow-shafted subspecies have engendered an orange version that lives east of the Rockies. Why are there so many eastern and western species of birds? I've wondered. Canadian biologist E. C. Pielou explains that during the Ice Age a cold, windswept desert covered the center of the continent. This barrier, she suggests, separated and isolated the ancestors of present-day birds, and over time the differences that characterize them evolved.

Where the morning sun strikes the rocks and foliage, vivid reds and greens become richer. Almost obscured by mist, deer browse in Horse Pasture Canyon, a grand enclosure that extends almost two miles from the main Dark Canyon. Such parks are generally less appealing to deer, which graze only certain grasses, than to the horses and cows that once found fodder here.

But today, without the heavy grazing, marsh grasses and rushes and sedges flourish in this high country swamp. Among them grow thickets of bushy alders, willows with yellow "pussies," and chokecherries whose heavy aroma fills the humid air. To attract bees and flies for optimal cross pollination, chokecherry flowers exude both sweet and rank smells. A couple of boggy ponds with marsh mallows in golden bloom gurgle into the creek. With their seeds well disseminated by waterfowl, many such wetland plants exhibit a remarkably broad distribution.

Moisture also determines what trees grow where. In the marsh spreading box elders prevail. On its edges, graceful firs and quaking aspen dominate the wetter slopes, while arched scrub oaks and towering ponderosas rule the drier ones. As a crack in a green-draped table rising above the desert, Dark Canyon presents ecological paradoxes. Mountain and desert species flourish side by side. Black bears tread the same paths as ringtails, mule deer browse near desert bighorns, and mountain bluebirds and horned toads stalk the same bugs.

Why can't we make this place our Walden, taming birds like Henry David did at his shack? There's a cozy cabin, a waterfall, a spring, wildlife, plus beauty every hour of day and night. Granted, there's no Concord two miles away when food runs out, but so what? Perhaps if we learned more about edible plants, we could survive. There are chokecherries, cattail roots, watercress, pinyon nuts, and acorns from scrub oaks. Perhaps the pioneers could have softened their hardships if they'd gathered more native plants and depended less on wagons and trains for foodstuffs like coffee, flour, and lard. Somehow the cowboy ethic of self-reliance didn't apply to food.

Nor are such yearnings for renewed natural connections unique to Anglos. Lakota medicine man Pete Catches expresses this same longing: "I want to withdraw further away from everything, to live like the ancient ones. . . . to become part of the woods. There the spirit has something for us to discover." Here Native American wisdom and modern therapy coalesce. In American culture we are just coming to understand the healing potential of nature. Its mysteries fascinate not merely because they dispel boredom with its deadening, dispiriting consequences, but because they allow us to live spontaneously without needing to force our attention. This alone could help to dispel the "dark melancholy" that Thoreau sought to transcend.

Before we leave the cabin, I visit the doorless outhouse. On the wall a toker—not a roper—has scrawled some classic lore: "Weed is the laxative of the mind. —A. Head '69." Dried rat pellets virtually cobble the floor. To insure maximum protection for their nests, pack rats have stuffed the facilities with sticks and leaves, barely leaving room at the top for new deposits. It feels good to drop a contribution on their abodes, I'm ashamed to confess.

Our way out seems simple enough: follow the drainage to the head of the canyon, then bushwhack up to the road at the Notch and follow it to Barbara's car, six miles distant. According to the map this requires a climb of thirteen hundred feet to the plateau. This isn't too bad, at least on paper. There are few trails and no footprints here in upper Dark Canyon. The area is remarkably pristine, the way things were when the earth was young and the West was still "a virgin land," at least compared to the ravished places the explorers left behind.

Grasses soft with dew tickle our calves along the trail. Zigzagging between the sharp branches of the giant sage bushes, I navigate with outstretched hands. My eyes scan the streaked slickrock and Barbara's hard, strong legs. Little do we know how much she'll need them today. After a couple more miles the gully tightens into the head of the drainage. A Western tanager, her pale yellow perfectly blended with gray, looks us over. Under the cover of protective coloration, female tanagers do not flush easily, especially when sitting on their nests. Still, how refreshing it is to be a curiosity, not a threat, where humans are not a regular presence.

Their leaves already turning brilliant scarlet, wild geraniums blush pink among smooth aspen trunks. This area must glow in the fall, when aspens glitter gold and orange against red rocks, when these rabbit brush flare into golden bloom. On these brilliant fall afternoons, flying grasshoppers will crackle their last, oblivious to impending snows.

Hunched under our packs, we lumber toward open slickrock. Already panting, I suggest stopping for a drink, a likely pretext to get a load off my back. Soon we're loaded up again. Exploring back and forth on the slick hogbacks, we hope to spot a route to the next level. Finally Barbara scrambles up a cleft, reaching for the next platform. Her fingernails dig in hard, but skid. Then I try. No traction. She wedges a log for me to stand on. No go. If only we had a rope to loop around that juniper trunk.

"Let's check that map again," she puffs. Just as before, the topographic map shows a route to the head of Dark Canyon. But there the canyon diverges into several drainages where we've probably taken the wrong finger. Even though the map clearly doesn't fit the territory, I find myself fighting the decision to turn back.

Although I'd rather not reveal it to a new companion, I still carry a stubborn streak of "Yankee efficiency" that Thoreau would have deplored. Whether on the trail or elsewhere in life, I hate to backtrack. This compulsion toward maximum output and minimal time costs me more than I realize. Ironically, I often invest more energy avoiding duplicated effort than it would take to do the haul again. Efficiency is a workplace value; it doesn't belong in nature. This mentality of efficiency no doubt greased the gears of the Industrial Revolution, but it's not a value I want to live by. Strange stuff can surface in the wilds, as Thoreau discovered when the raw wilderness of Maine nearly shook him out of his britches.

When we reach a fork, we probe one more drainage. This one ends at a chute, but nearby stands a silver fir, a very symmetrical, grayish-green true fir that grows where it finds ample moisture. A branch extends over a dip in the wall; it might provide a way up there. A self-confessed tree hugger, I don't hesitate to climb it even if its sap sticks my fingers together. But this branch is another dead end, for it won't hold me. As clouds and cold air move in, the fingers of Dark Canyon seem to claw desperately at the saddle we seek, above and beyond.

Relieved to follow familiar footsteps, we decide to double back to Horse Pasture Canyon where the map shows another way out. As I breathe deeply under my load, my mind exhales the frustration caused by my attachment to goals. If I want to experience and grow, I won't seek the easy, familiar cow path that we followed in.

As we devour cheese and saltine packets, ravenously licking the containers, an odd grinding comes from a small cottonwood. Saying nothing, we tiptoe away from the tree, then point simultaneously to a young porcupine near the end of a branch. Its small black eyes are shy and unexcited. Oblivious to us, the porky chews one twig after another. Protected as they are, porcupines don't spend much time worrying about predators.

After another look at our soggy map, we trudge past the cabin. This time it's the old shed that promises new bygones. Mouse-chewed feed bags and saddle pads hang on rain-stained boards for riders who won't return. Rusty horseshoes rest on their rack, ready to go. A hitching post still swings from the rafters, awaiting the horse and cobbler. On our way out we amble along the bleached fence, passing beneath the simple log entry to the ranch. In addition to its better-known charms, southern Utah remains rich in historic cowboy sites: corrals, caves, shacks, even stone cabins built right into over-hangs. It's sometimes difficult, however, to reconcile my interest in historic cowboys with my revulsion against the overgrazing they initiated.

Broad-rooted ponderosas, their long needles streaming and whirring with the breeze, grace Horse Pasture Canyon. "Of all the pines," John Muir remarked, "this one gives the finest music to the winds." The most common pine in the

West, these glorious trees take their name from their heavy, "ponderous" wood. Ponderosas also sustain tassel-eared gray squirrels that feast on their seeds, buds, and vanilla-scented inner bark. Few mammals feed so exclusively on a single plant species. These squirrels avoid trees with more resin, which can be toxic. This avoidance may contribute to natural selection, for those ponderosas with the most resin thus become more likely to reproduce.

Soon sedges bend and squeak as the bog squashes beneath our feet. If we don't have better luck finding this trail out, we'll have to head out the way we came in, which would mean another encounter with those cows.

Eureka! A scar marks a trail. After a scramble up its grade, we grab for roots to make headway on loose sand, then ignore our senses and gut it out, blinded by sweat. I stumble into a box elder that clicks and buzzes: quarter-inch leafhoppers jump in all directions, especially toward my scalp. This tree definitely isn't a climber.

When we finally reach the road, we dump our packs into the dust. Free at last, we feel light enough to soar. The plateau drops off on both sides at the Notch, the narrow swayback that separates the many-fingered head of Dark Canyon from that of Arch Canyon which drains toward the San Juan River. At the Notch we stop to look for the trail we couldn't find. There, directly below us, stands a trail marker but no trail. Hundreds of trails disappear each year because National Forest priorities foster lumbering, grazing, and mining over hiking.

Silhouetted on a rise, coyotes yip at the wind as it ruffles their fur. The howls flush wild turkeys from a thicket where they're feasting on acorns. The turkeys gobble in distress, their legs becoming a blur. I wonder how many coyotes will survive the winter when government gunners mount an offensive, often from airplanes, at the behest of public lands ranchers.

During our next rest, the bushes rustle. A black hulk lumbers, cracking twigs. We think bear. A large head appears, but it's not a bear's. Instead a monstrous bull stares at us, then lowers his head. We slip our packs over one shoulder and backpedal, glad that both packs are blue, not red. We make a quick detour, keeping one eye on the bull, the other on large trunks we could hide behind in case he gets ornery.

When at last we find our car, we're a couple of hurtin', hobblin' buckaroos. Our packs have extracted penance for our urban sloth. If we'd dragged crosses, our flesh couldn't feel any more mortified. My old obsession with backtracking now seems silly. After all, many paths in life lead nowhere, and dead ends occasion new discoveries. Why, if we hadn't had to turn around, we would have missed that bull.

This was a great trek, a fine mixture of history and natural history. My only regret is that, with all the signs of human activity around, I clung to a civilized consciousness. I want to become more animal out there, freeing not just the child inside but the beast as well. That will require more than two days. Just as the center line can linger in our eye long after we park, we carry cultural baggage with us. Next time, for starters, I'll leave *Walden* behind.

Interlude

A Rejuvenating Plunge

Tired of hiker chitchat, I head for the water's edge. Never mind that this is a mountain lake; a distracted mind and sweaty body scream for relief. Enough talk, too much heat. My clothes pile up on the grass. Achy, puffy feet dodge alpine thistles as they tiptoe toward the shore.

Once I wade in, the chill hits. A cloud eclipses the sun. While the other hikers jabber, I look first at my own reflection, then for a pebble to skip. When a gust blasts my clammy back, my arms clutch my bare chest. I toe the bottom, stalling as minnows nibble my goosefleshed calves, then I wade on tiptoes, wobbling on slippery stones as waves slap my thighs.

I dive in, and my torso stings hot. Surfacing, I blow out frustration and synch my breaths to my strokes, slapping and kicking for warmth. The shock wanes. Froth sloshes in my ears to drown out the chatter from shore. My fingers send bubbles streaming along my arms, past my belly, then down my legs. My torso pitches and rolls like a schooner. Bands of sunlight fan out like a corolla to energize my shadow as it crawls the bottom. Seduced by the ever-deepening blue, I sail over the snags and relish the warmer surface water. Stroke after stroke, I drum a cadence for my lungs.

But a wave floods my mouth just as I inhale. I choke, breaking the rhythm. More waves slap my face. I try to catch a breath. Fear creeps in like the cold as my feet stab down, only to find no bottom. My chest tightens, my arms tense up. I take short gasps. Spit water. No air. I run on raw survival instinct. OK, I've swum too far from shore. Keep lungs filled—more buoyancy. Which way back? Eyes scan the blurry shore. Nothing but trees. Oh, there's somebody staring into the shallows. Too far to hear a yell for help.

Christ! Am I about to drown? Got to get moving. The cold will finish me here. One good breath, two, three. Got to keep within my breaths. Hands and feet slap the water like beaver tails, this time in terror. My arms weaken and my legs take over.

There's still no bottom in sight, though the water begins to warm. Only another fifty yards. Those waves are gone. Gusts still drive spray into my eyes. Only a few dozen more strokes—assuming I don't stray off course. Gotta gut it out.

Finally a gravel bar. I could put my feet down here but I won't. Better to finish strong. My rubbery arms can still dog paddle into the bath water that, only minutes ago, felt so frigid. My nails claw the cobbles. This time I let these rocks tickle my screaming thighs. Fingerling brookies flash silver, darting for cover. Only six inches below my dazed eyes, blurred stones glide by like dots in an impressionist painting. Golds, oranges, and chocolate browns waver in the crystalline light.

For a moment I linger, elbows on the rocks, chest heaving, legs tingling, spent, shamed, but ecstatic to be alive. I savor the black muck that squirts between my fingers, mindful of my kinship with this ooze. Then I crawl out, exposed, very pink yet now beyond embarrassment. Shivering enough to rattle teeth, I grab handfuls of grass to pat myself dry.

It was pretty foolish—all right, reckless—to let myself be lured into deep water. New eyes behold the greenness of thistles that I almost didn't see again. New ears full of water pulse to drown out human voices.

5.

Flowing with the Green

I was trying to concentrate all my strength on my ardent
desire to break through the crust of the mind and penetrate
to the dark and dangerous channel down which each
human drop is carried to mingle with the ocean.

—Nikos Kazantzakis, *Zorba the Greek*

After rains throughout the Rockies, the snow pack is melting fast. Here in Utah the Green River is rolling at eleven thousand cubic feet per second, perfect for an expansive float on the runoff. Barbara and I follow dancing flashlight beams toward the park ranger's show. His slides look like home movies, but he spins yarns like a mountain man. In the nineteenth century, he tells us, the hardy voyageurs who trapped beaver over a vast area often met here in Green River, Utah. Then, donning a beaver hat, he lets out a whoop that scares the kids.

This guy's a good showman, but he doesn't mention that trapping was destructive to ecosystems over a vast area. Much as gold would later draw prospectors westward, furs lured trappers to Wyoming and Utah as early as 1815, only a decade after Lewis and Clark first explored Western rivers. By the 1820s beaver barons such as Jim Bridger and General William Ashley had already trapped out whole watersheds. The rage for beaver hats peaked during the 1820s and 1830s, but beavers faced trap and gun long after the fashions changed. By 1900 these remarkable rodents survived only in isolated pockets of the West. However, beavers have made a comeback, and they've even become a pest in some cities where they drop ornamental trees.

At sunrise Barbara and I inspect the river while we inflate our raft, an eleven-foot Sevlor that we'll paddle together. Cottonwood puffballs fleck the brown water where large flannel-mouthed suckers, their peach-colored mouths against the current, seine the surface for insects.

As we shove off, the willows on the bank glide by like a slow-motion shot in a film. Her tresses uplifted, Barbara sits like a mermaid at the bow. A pair of stately Canada geese, their goslings' yellow heads blending into their greenish down, paddle along the bank and honk softly at us. As campground noises fade into the past, we peak our paddles to drip, drip, drip while tiny waves thunk on the bow and rock our boat. The girders of the Denver and Rio Grande trestle, first built in the 1880s, clank as they expand in the sun. Just below this bridge, in 1889, Robert B. Stanton's railroad survey team set off looking for rail routes to California.

Allowing four days to float sixty-eight miles, we'll be spending six or so hours a day in the flow. Though it drops only about a foot or two each mile, the river rolls along at three miles an hour, swirling and eddying, its riffles now roused into chop by strong gusts that almost blow us into the bank. Breaths synchronized to strokes, I paddle myself into a meditative state, beyond time and place. When my eyes open, it seems strange to be on the river.

Not far above water level the yellow and orange rocks resemble those of Yellowstone. On his famous first expedition in 1869, Major John Wesley Powell described these as "deposited by mineral springs that at one time must have existed here, but which are no longer flowing." Sheltered beneath the mineral-stained rocks, we manage to convince each other that we'd better rest here before facing that headwind again.

Nor is this the most scenic stretch of river. Beyond sand dunes and a large butte, the San Rafael River enters on the right, overgrown with tamarisk. Major Powell, who showed a keen interest in archaeology, found this spot intriguing as "a frequent resort for Indians. . . . Flint chips are strewn over the ground in great profusion, and the trails are well worn." This tributary drains the San Raphael Swell, a magnificent sky island sculpted from the sandstones of the Colorado Plateau.

Decades later, at this very spot, the steam launch *Major Powell* sheered its propeller on a boulder. According to river historian Roy Webb, this tour boat scheme was concocted by B. S. Ross and his wealthy backers. When the *Major Powell* floundered on its maiden voyage (with its backers aboard, no less), the embarrassed Ross commandeered rowboats and continued the trip to the confluence of the Colorado and the Green, where he proposed building a tourist hotel. Like Stanton's expedition, these entrepreneurs underestimated the resistance of these river canyons to industrial development.

A sputtering tractor heralds Ruby Ranch, where the Green begins its long and final cut through the red rock. Gradually the mud-and-gravel banks rise into rounded sandstone mounds that resemble Navajo hogans. Salts encrust the rock and leave it brick red where rainwater has washed them away. For a better look we beach at Bull Bottom, where brown-speckled eggs litter the tawny sand and nests dot a guano-streaked overhang. The cliff swallows alight and fan their tail feathers, prompting a chorus of cheeps. Other swallows swoop, veering away at the last instant, then flash their rust-colored rumps. Magpies jaw and jabber. Then all falls silent except for the whispers from the great stream: so much mass in motion, so little sound.

Now over five hundred feet wide, the river slides beneath our raft. Reflections animate the shaded smoothness, where bands of brown, green, orange, and blue fuse together. Upwelling currents become boils of liquid amber, then eddies dimple the whirlpools of molten mud. For miles we float languorously, entranced by the dreamy light and solitude, the silence broken only by muffled splashes from the boils and the trills of distant songbirds.

Leaving a V-shaped wake as it carries a branch in its mouth, a huge beaver slaps us from our reveries. In times past wildlife abounded on the Green River. Even after several decades of trapping, Powell's boatman Jack Sumner reported that "the whole river . . . seemed to be swarming with beaver. We shot several with our pistols as they bobbed up near our boats. I had the good fortune to get two otter out of a bunch of five as they swam past puffing and spitting like a whole nest of tom cats." That these explorers so shamelessly touted their killings suggests the habit of random violence in the Wild West.

How markedly these frontiersmen's attitudes differed from those held by the Western Indians they replaced. Mrs. Medicine Bull, for instance, recounts a Cheyenne legend:

> There's a great pole somewhere, a mighty trunk similar to the sacred sun dance pole. . . . The Great White Grandfather Beaver of the North is gnawing at that pole. He has been gnawing at the bottom of it for ages and ages. More than half of the pole has already been gnawed through. When the Great White Beaver of the North gets angry, he gnaws faster and more furiously. Once he has gnawed through, the pole will topple, and the earth will crash into a bottomless nothing. That will be the end of the people, of everything. . . . So we are careful not to make the Beaver angry.

This reverence for life, anchored in a belief that all of creation carries a spiritual dimension, helped native people to survive for thousands of years on very limited resources. Only in recent years has the public started to comprehend the magnitude of what Euro-Americans lost when they killed Indians—rather than learning from their long experience on this continent.

In the distance Boy Scouts yell and listen to their echoes. Perhaps for lack of anything else to do, they have amassed driftwood for a blazing bonfire. Even though they're baking in the sun, they heave more wood on their already towering inferno. Far from the controlled conditions of home, suddenly confronted with raw rock and primeval river, they may hope to assert some masculine control to keep the wilderness at bay. It's difficult to imagine the Girl Scouts building such a bonfire.

With daylight waning, we scan the banks for a campsite. Bursts of paddling twice fail to propel us across the current in time to beach at likely spots. Beneath its placid surface the river runs with great power. Finally we grab a willow and scramble up a stony bank, nearly starting a small avalanche. Famished, we tear into our grub and chase down every last grain of rice. Our stove roars like a blowtorch; when it quits, I wonder how noise can make silence seem strange, unnatural, and how much we become inured to noise in our daily lives. City noises—from screaming sirens to thundering speakers—all seem to dull us. But if industrial civilization requires desensitization, silence refreshes and relaxes.

Tonight, on Midsummer's Eve, the late day light is particularly spellbinding. It retreats up the canyon walls, highlighting rocks and trees that in other lights I would have missed. Atop the walls wind-rounded junipers stand silhouetted against the orange-and-violet sunset. As dusk falls softly, it pools beneath the scrub oak and slowly fuzzes outlines even where cliff meets sky. Along the river the twilight air turns damp and musty, full of funky patchouli. As the chirping of birds fades into the rustling of rodents, clock time becomes meaningless, replaced by the rhythms of day and night, of the restless river and the flowing mind. We twitch when a ground squirrel drops down on our tarp.

Between curving canyon walls that channel a starry stream, the Milky Way now illumines a sluice of sky. The distant heavens appear just beyond reach and society seems far away. Of course this is a double delusion. The Milky Way is so vast that we see one hundred billion stars blended into a faint cloud of light. Rivers, on the other hand, carry so many more people every year that they can seem like highways.

However close society encroaches, though, a night full of stars can expand the mind. Imprisoned but coming more fully alive, Albert Camus's character in *The Stranger* gazes into the swath of sky that's visible from his cell. For him simply watching the heavens makes life worthwhile. Just before he's executed, he begins to embrace "the benign indifference of the universe."

Because distant stars hold no apparent human significance, however, "benign indifference" has not met psychological needs for connection and meaning. Seeking relatedness, the ancient Chinese called the Milky Way "the little sister of the rainbow." For similar reasons the Greeks created constellations to represent gods, heroes, and animals. Extending traditional sun-as-male/moon-as-female mythologies, St. Francis personalized the cosmos as "brother sun, sister moon." In recent decades countercultural tides have turned to astrology's depiction of the sky in familiar terms. Projecting human qualities on nature makes us feel less insignificant or alone.

Feeling lost in a meaningless universe, people may continue to search for a creator's signature, an assurance that "somebody up there designed all this." But the experience of seekers, astronauts, shamans, and mystics, many of them in non-Western cultures, would seem to suggest that spiritual connections are possible without having to discover the creator. Although painting a human face on the spiritual presence may heighten our sense of relatedness to the divine, such personified gods may also distort the sacred mysteries of the universe.

In the morning the air is damp along the bank. Chilled, I shift from sandal to sandal, keeping my knees together. As sunlight creeps down the wall, though, the reflected warmth toasts my legs. Then the sun hits. In only a few minutes we go from hunching, coffee cupped in hands, to basking in our swimsuits. Once launched, we enter a swirling eddy where a huge catfish, white belly up, bobs in the flotsam, fodder for ravens, vultures, eagles, or other fish. Much of the canyon lingers in shadow. The liquid trill of a solitary canyon wren cascades down the sheer streaked walls. In this stark country you can hear a single bird or see a single plant across great distances. Rain forests are biological treasures, but they're so densely packed that it's tough to focus on an individual plant or animal.

When the sun surmounts the rim, we take turns floating. My eyes peer far downstream, vigilant for snags, while Barbara sighs in delight when a cool upsurge boils up, glassing the river's surface. Lips at water level, she seems oblivious to the yellowish tamarisk pollen and pulverized wood. When my turn comes, though, at first I find myself repulsed by the river scum. But total body stimulation sweeps me out of my mind, toward oneness with the flow.

Around a great bend we hug the bank to spot Trin Alcove, a confluence of three canyons. As thickets of feathery tamarisk and scrub oak glide by, seeming to flow upstream, voices drift from a small lagoon. The Boy Scouts gawk as our raft appears, mirage-like, revealing only a mermaid in a bikini. "That bearded guy with her must have drowned or hiked out," they probably fantasize. Behind the raft, unseen, I imagine their faces when they see me emerge from the river, muddy as a hippo. As I fumble to untie my swimsuit from the moving raft, Barbara keeps right on paddling. OK, the scouts be damned. I toddle out of the water wearing just a smile.

In the winding glen trumpet-like scarlet, gilia blaze against the buff sand. Long runners from giant tufted cane shoot over the cracked mud, sending down spikes every foot or so. Moments later a rock panel rivets my gaze. Highlighted by the noonday sun, conchoidal fracture lines radiate out like shock waves from where a huge hunk of sandstone broke away. On this fresh blond surface, water streaks range from muted cinnamon to nutmeg to chestnut—a full spectrum of delicious browns. These streaks often fan or fade out, altering their shading where the grain shifts direction. Like much of the red rock in the Southwest, this Navajo sandstone is sweepingly cross-bedded, laid down as dunes and later cemented with lime beneath a shallow sea.

In this acoustic amphitheater clicks seem to come from all directions. Camera in hand, another river runner waits until the sun highlights more grain in the stone. Poised to photograph, will she observe the rock more carefully? Possibly. Or will she frame her vision with 3 x 5 rectangles and preoccupy her mind with what settings to use? Photography is a great medium, a wonderful intersection of art and science. When most of us carry a camera, however, we miss more than we capture. Yet I also have my pad of paper handy and sometimes miss what transpires while I scribble notes.

Exploring a second canyon, we hike up the main wash. After two hundred yards, sweat stings my eyes as the cliffs writhe in the heat. We settle for a drip beneath an overhang to fill our canteens, drop by drop. With our shirts soaked we can face a blast of light that glares off quartz grains to bleach the sandstone. I hug the shady side and splash through puddles, hat lifted to keep a cool head.

Sheer, smooth walls tower several hundred feet above the canyon's floor. Here, where the sun seldom penetrates, the air suddenly turns cool and moist. As I wade into this inner sanctuary, I can even see my breath. Each drip resounds in our ears. The canyon ends with a dark twenty-foot hanging garden draped with stalactites and matted with moss. My arm sinks into the wet tapestry. Then a whir fills the cool grotto as a hummingbird hovers, twists its body for a look around, and shoots straight up, leaving shrill squeaks behind. Barbara and I look at each other as if we'd seen an eagle.

About eight feet up, a brown hellebore orchid embellishes the mat of moss. To see it better I begin to wedge myself against the sides of the slippery chute until Barbara intervenes: "Are you trying to break a bone?" She's right; this is no place to break a leg. I jump down with a hollow crunch on the gravel.

Back at the river a scrap of paper hangs on our cooler: "Have one on the troop." Tucked inside, chilling on fresh ice, gleam two cans of beer that we swig in the shade. As I try to push off the bank, my feet stick in the muck deposited in this backwater. Meanwhile the current begins to catch the raft, dragging me forward into the churned water. At last I'm able to flounder over the tube, smeared with clay. No one takes the helm. For miles the red rock, blue sky, and puffy clouds spin by as the relaxed river buoys us along, aided by a boost from the beer.

Our second campsite is buggy. A sound like a stiff brush rasping a screen makes me shiver; the deep buzz of this June bug is unsettling. Red ants find our tent and small flies cloud our faces, relenting only when Barbara lights a fire. As if our skins were not already abused enough from sun, wind, water, and silt, now we add bug bites, heat, and smoke. Barbara hikes upcanyon, returning with some fossil crinoids and brachiopods she's found. Though they're often mistaken for plants, crinoids were actually the stems of sea lilies, the ancient relatives of sea urchins. The equally common brachiopods were bivalves resembling clams.

The first stars wink through the clicking leaves of the cottonwood. When Einstein asked, "Is the universe friendly?" he indulged fond hopes, but right now it seems to whisper, "Yes." On balmy summer evenings like this I feel beyond time, free of worries about change, loss, or death. Blessed and blissed.

The following morning I awake early, feeling clammy under just a grungy sheet. Propping myself up on one elbow, I note a fuzzy jet contrail streaking the sky. Humidity. The steam droplets rise fully two feet above our coffee mugs, looking granular in the hazy sunshine.

Where there are bugs, there are birds. Bullock's orioles, their flights looping like telephone wires, flash their yellow and orange plumage. A Western kingbird, its gray throat blended into a lemon-yellow breast, perches high on a dead branch to watch for flies. A rufous-sided towhee with a brick-red breast rummages in the dead leaves. Such diversions make it pleasant to scrub oatmeal from a pan or to swab mud from a raft. Under the right working conditions, even dirty jobs can be a delight.

After our launch we watch for a drainage. The mouth of Ten Mile Canyon appears, as side drainages so often do, like a pond behind a muddy sandbar. The muck sucks on our paddles as we push into a backwater enclosed by tamarisks and willows. Dragonflies bank and circle. Jointed walls of Wingate sandstone, one of the great cliff formers in this country, tower far above. Well-watered and protected from grazing, here the feathery-gray sage grows five feet high. Where this stagnant creek narrows, tufted cane, coyote willows, grotesque hackberries, and droopy box elders fashion a jungle in the desert.

As my paddle strikes bottom again, I slip over the side to propel us paddle-wheeler style, though I don't enjoy the muck and scum. Kicking the boat from behind offers a duck's-eye view of banks inscribed with animal prints. A splash startles me; the raft rocks. A beaver chute of smooth mud leads to a hole in the bank just a few feet from my face. When this drainage ends, will I have a tooth-to-tooth encounter with this fellow? Beavers run up to seventy-five pounds and flash bigger buckies than we river rats do. As I stagger forward, freckled with duckweed, the bottom bubbles and I drop into quicksand. Calm as usual, Barbara tosses me a paddle. I flail for a moment before finally extricating myself, minus my thongs. Shaken by the experience, I review what I know about quicksand. It requires fine, saturated sand with enough water seeping through it to lift the grains into suspension. All this science and reason, however, doesn't entirely quiet nerves on full adrenaline alert.

Once the rope is looped around a willow clump, we walk on firm, furrowed sand whose cracked, shiny surface reveals that water has flowed here recently. Since then, downcanyon gusts have left the coarse sugary sand at the top and deposited the finer, darker grains at the bottom of each rill. Windblown milkvetch with bulging pink pods etches crescents on the smoother sand. A pastel-green leafhopper and two chartreuse grasshoppers cling to a red-striped tumbleweed.

On even more desolate gravel nearby, strange bottle plants are already brown. Discovered by explorer John Frémont in 1844, these members of the buckwheat family exhibit a classic adaptation for storing scarce water. The Paiutes showed pioneers how to use these botanical curiosities to quench thirst and provide food. After shooting out a rosette of leaves at ground level, the plant's several stems elongate into "bottles." From these, spindly branches lead on to smaller "bottle stoppers," then on to tiny yellow flowers that last but a short time. Deserts speak of both extreme endurance and extreme transience.

While thunder rumbles in the distance, hot sand scorches our bare feet. We spot a spreading cottonwood and trot for relief in the shade. Above us, amid the large leaves perforated by small holes, the cotton pods burst to release filaments that blur the line between fuzzy tree and hazy sky. Large scarabs with black legs like grappling hooks clasp the leaves. These amber-winged goldsmith beetles, the gold bugs in Edgar Alan Poe's celebrated story, hang on twigs in copulating pairs. Others buzz slowly around the branches. It is now clear what is perforating the cottonwood leaves. In fact, hungry goldsmith beetles can strip a tree entirely of its foliage. Because desert plants are so infrequent, they often get hit hard by bugs.

Nevertheless, life flourishes amid this heat, rock, and sand. Squat plants engender star-shaped flowers only one-sixteenth inch across. These tiny blooms range from yellow-green to creamy tan, depending on how long they've been open. Beside these, bleached sepals remain fully open in rigor mortis, dried before they could wilt. In reality, the desert is no harsher an environment than any other for species that are adapted to it, but because of its dryness the relics of life remain, decaying very slowly. The desert keeps death visible, reminding us to seize the moment.

We hear distant thunder. Above the mesa, black thunderheads loom like jellyfish with tentacles bowing beneath them. Windblown curtains of rain, these verga billow but don't reach the ground—at least we hope they don't. A downcanyon draft rattles the cottonwood, dislodging puffs from its branches, and then brings a brief whiff of ozone. Rain. The rumbles were miles away but as we set out for the raft, scum laps at our heels.

We glance at each other, neither of us willing to utter "flash flood." Down the drainage races liquid mud. Dried leaves, pinyon nuts, and shriveled juniper berries ride the scummy crest as tumbleweeds and broken branches snag on willows. We watch the depth of the slurry, looking back for the proverbial wall of water. Our feet catch on roots, sticks, stones, and even clumps of grasses that are now skidding along. Rain craters our bare bodies, leaving rings of sand. Without saying anything, we both wonder what is happening to the raft. It could be inconvenient to float the river with nothing but air-filled canteens for life preservers. The thunder claps again. Clouds are bursting somewhere. We keep running.

Our raft is tugging at its tether. Oblivious to thorny brush, we drag it to higher ground, dig out our raincoats, and glance at the skies. The torrent of pink froth abates with no real flash at all, leaving a brick-red sludge. The creek now pours over the sandbar, marbling the beige water of the Green. Far above, steaming red cliffs jut angrily against scowling skies ready to strike with lightning bolts. Pink torrents spout from hanging chutes, splatter into bowls, then plummet farther down to raise more spray. Light on the retreating mists paints a rainbow that arches from rim to rim. Its almost neon colors pulsate while we shiver, rapt at the spectacle.

The Green seems aptly named, not because of its own color but because its silts grow such lush greenery. Compared to those of its sister stream, the Colorado, its banks are indeed greener. When the two great rivers reach their confluence, the Green has come much farther, allowing it to pick up more sediment. Additionally, the Green cuts through the limestones, siltstones, and shales of Desolation and Gray canyons which erode into clay and mud, while the Colorado courses through sandstones for most of the two hundred miles above its confluence with the Green. Both rivers carry silt loads that, gallon for gallon, far exceed those of the muddy Mississippi.

Soon the sun reappears and the runoff subsides. We hope to explore Bowknot Bend, the great gooseneck of Labyrinth Canyon where the river has carved a hairpin isthmus and a nearly circular "island" miles in diameter. In this deeply entrenched meander the Green flows about nine miles while progressing less than one mile downstream.

Watchful for the isthmus, we soon we spot a historic "billboard" where "Launch Margarite 1909" is painted on the cliff. During the early 1900s this stern-wheeled steamer plied the waterways between the towns of Green River and Moab, heading upstream on the Colorado. Astounded by the geology, intoxicated by the heat, and distracted by the billboard, we somehow sail right by the saddle between the goosenecks. Because the river arcs so gently, it's difficult to tell that we're completing a grand circle.

Below Bowknot Bend, the current slows slightly where Labyrinth Canyon gives way to Stillwater. Major Powell, who named both, described these rocks eloquently:

> We glide along through a strange, weird, grand region. The landscape everywhere . . . is of rock—ten thousand strangely carved forms; rocks everywhere, and no vegetation, no soil, no sand. In long gentle curves the river winds about these rocks . . . all highly colored—buff, gray, red, brown, and chocolate, never lichened, never moss-covered, but bare and often polished.

The rocks here are more varied in color, less rounded, and more broken in their layering than the Navajo sandstones upstream. These tiered walls rise over a thousand feet, separated by a quarter mile of pristine air. Whereas the Navajo sandstones form slickrock domes, here the reddish-brown Wingate sandstones and the purplish-red Kayenta formation fracture into spires and sheer cliffs.

Contrary to Powell's description, though, this area is hardly devoid of vegetation. Willows and cane wall the banks along with tamarisks, which admittedly were not here in Powell's time. Behind this barrier of greenery grows an occasional cottonwood. Higher up, yuccas, cactus, pinyons, and junipers dot the slickrock. No doubt Powell's eye compared the Southwest to the more verdant areas he was familiar with back east. Perhaps his inability to perceive the flora suggests that in the absence of what he was used to, he saw "less" as "none." Or perhaps his experience provides a parable of over compensation, much like the tendency to see pink after removing green sunglasses. Whatever the psychology, it often does take newcomers some time to adjust to the more spotty greenery of the canyon country.

Greenish-gray Chinle shale now surfaces along the river. Another characteristic formation of the Southwest, the Chinle is famous for three things: spectacular colors that peak in the Painted Desert of Arizona, petrified trees found at nearby Petrified Forest and elsewhere, and uranium ore. In ancient stream environments, decaying plants often accumulated at a bend in a river where, over time, uranium oxide gradually replaced the minerals in the plant matter. As a result prospectors have located whole petrified logs of uranium ore. Whether or not humanity needs any more uranium today, during the 1950s the price rose high enough to encourage exploration even in such remote areas as this.

Emerging from the seven-mile meander around the

Bowknot, we float under a memento of this era: a rusty mining bucket hanging, doors open, from a cable that stretches across the river. This apparatus would have ferried uranium ore to the rough road that still incises the east bank. Prospectors' tracks not only scarred the canyons of southeast Utah, they opened the area to both responsible four-wheeling and to off-road vehicle abuse.

Our river mileage logged for the day, we beach on a sand spit. Since there's nothing we can tie to, Barbara uses a trench shovel to bury a log around which I've wound our line. Duty done, we recline on the sandbank, protected from the wind. What's a little sand in the scalp when you've got it everywhere else? Soon, too soon, a flotilla of canoes appears, their paddle flashes visible a mile upstream. To our dismay, all six canoes land on the spit. Trying to puff up like swans defending turf, we stand as tall as we can. We contend that there are other good campsites. The canoeists counter that their children are tired and can't paddle any farther. We trudge back, deflated after our failed puffery.

As the intruders set up, grainy gusts flap their tents. When the sand stings our legs, we reluctantly push off and "sail away." A mile downriver we spot another site but together the wind and the current skid us sideways. "Pull, pull, pull," I shout as we synch our strokes. But the current veers toward midstream. "Pull, goddammit, we're going to miss this one!" I snap. With the rope in one hand I stretch, hip on the tube, to grab the rock. Legs splayed so as not to get pulled out of the boat, I gradually work us out of the current. We tie up in the eddy. Although the wind doesn't relent, at least we're not facing the sandstorm that still peppers those other folks.

The next morning we set out early, before the heat hits.

Soon we arrive at the mouth of Hell Roaring Canyon, a place of Sonoran/Mojave desert flora. Sage and four-wing saltbush, its leaves bleached almost colorless by the sunlight, inhabit the alluvial fan. Shadscale and greasewood dominate the talus slopes of broken Chinle shale. The now-rare Isely milkvetch once grew on these gray clays, but during the 1950s it didn't take prospectors long to realize that the selenium-laced soils favored by these legumes often also contained uranium. Many plants were lost in the scramble. As the sun pounds palpably, we hike up a wash rutted by trail bikes. Will the Isely milkvetch survive the mining onslaught of the 1950s only to perish beneath the aggressive treads of the 1990s?

But these signs of man are not the ones we're seeking. A couple hundred yards up, "D. Julien 1836 3 Mai" appears alongside what looks like a sailboat. Regarded as the first European to navigate the Green, Denis Julien landed here thirty-three years before Powell. At first this date seems odd, since another inscription left by this French trapper twenty miles upriver reads "16 Mai." We rafters move with the current, but trappers in long canoes had to paddle upstream, loaded with furs, no less. Today we often underestimate the sheer physical exertions endured by early explorers, trappers, and settlers.

On the way back we tread over the deeply-cracked river mud, a perfect habitat for toads and miniature tamarisk seedlings. After pushing off again, Barbara basks in the boat while I bob in the river. A giant tumbleweed spins by, glistening like a water wheel. By now a faceful of dead gnats and froth seems normal. My inhibitions dissolve as I roll over and spout river at the sky. Leading with my feet, I let the flow whirl me as it will. My flesh sings to celebrate this triumph over an aversion that had limited my relationship with the natural world.

A car's horn suddenly pierces the quiet. Looking toward shore, we see what seems like a desert junkyard below the slash that criss-crosses the bluff. Sunlight glints off windshields at the parking lot. This is only our fourth day out, and vehicles already seem like spaceships from another planet. We've been away long enough to feel some reentry shock.

From the stern I frog kick the raft burst by burst toward the well-tramped bank. Swigging a tepid beer under a great cottonwood, we gaze into a mosaic of brown water, orange rock, green foliage, and blue sky, its tiles dancing in fluid, flickering shapes. Ordinarily separate entities—the me and the not-me, the heart, the mind, the body—all dissolve into one. As naturalist John Muir rhapsodized, "the rivers flow not past but through us, thrilling, tingling, vibrating every fiber and cell."

Floating is a good way to go. Exciting as it is, running rapids often turns rivers and rocks into antagonists. Besides, whitewater rafting has become so popular that to find solitude it's best to do a flat-water stretch where two people can float for miles, fully immersed in the flow.

Interlude

The Lemon that Almost Exploded

After setting up our tent I try to light our Coleman Peak I, the one that purred so nicely in the store. Being unfamiliar with this stove, I follow directions slavishly. Despite this special handling, the small stove spits and sputters, flickers orange, and goes out. Streaks of gasoline trickle below the burner. Reluctantly I pump it again, as directed, but this time there's no hiss at all. Only more trickles.

Caution warns that I should quit, but hunger prevails: our meals depend on this stove. Taking precautions, I light it and jump back. First it flares, then it goes poof. A flame jets out like a blowtorch. Will the blessed thing blow up, frying me? Is it going to start a brush fire? Barbara dumps a bucket of water on the stove, which refuses to go out. Then, crouched like a demolitions expert, she creeps up and heaves on handfuls of sand. This knocks down the flame but doesn't extinguish the incorrigible, inflammable Peak I.

Next Barbara gets another idea—knock the thing over and bury it. "Get back, that thing's a bomb!" I yell. Undaunted, she kicks the stove over and snuffs it out like a cigarette. We stare at each other in shock—and relief.

True, the damn thing didn't explode, but now we have no way to cook food or even boil water. Sure, somebody's got to get the lemons, but now what? We're six miles into the backcountry without a stove and it's getting dark. We've got three choices: start a small fire, breaking park rules; eat fruit tonight, collecting water from a seep and hiking out tomorrow; or borrow a stove.

A couple from Yakima, Washington, is just finishing their supper. They lend us their Peak I—of all stoves. When they turn it way down and hand it to me, all lit and ready to go, I carry it at arm's length, head turned away. We go to the wilderness to leave humanity behind, but when things go wrong, helpers are handy to have around.

6.

From Slickrock to Bedrock: Tested on the Dolores

To stick your hands into the river is to feel the cords that bind the earth together in one piece.

—Barry Lopez, *River Notes*

The Dolores—"the river of sorrows"—is running high, and none of us has ever faced real whitewater. Visions of *Deliverance* haunt my mind.

Before we can launch, we've got to ready our rigs. For John and Lori it's just a matter of packing a whitewater canoe and slipping a few extras into our raft. For Barbara and me, though, the task proves more difficult. The rowing frame for our new raft refuses to go together as it did in our back yard. As the sun pounds down, man and metal both turn more stubborn. Things aren't going as planned, and this doesn't bode well, especially before a first voyage. Once the frame finally does come together, Barbara pops a bottle of cheap champagne to "pagan" the *Canyon Wren*, a yellow, silver, and blue boat that will carry us far.

This raft actually floats! Life offers few better experiences than trying out a new toy. Riffles undulate the rubber floor as my muddy toes grip the shiny frame. When I spin us, amazed at the boat's maneuverability, pebbly clouds wheel by like stars in a planetarium. I lean into strokes and frustrations dissolve into the distance.

The signs of human disturbances, however, take more time to fade away. An abandoned mine, a relic of the uranium frenzy along the Uravan Belt, still disfigures a hillside. Where an oil rig burns gas and belches sulfurous fumes, startled cows snort and charge into the brush.

At a flow of two thousand cubic feet a second, the current runs cold and clean as it races toward pristine wilderness. Soon the river enters a sandstone canyon where cliff swallows nest under overhangs, then it meanders into one of the Colorado Plateau's many salt valleys. In eons past an uplift of salt has buckled the sandstone and shale once lying above it, making them more vulnerable to erosion. Here a wide, park-like Gypsum Valley affords spectacular views of peaks still packed with snow.

While a bird scolds us from the riverside brush, another sound grabs our attention: Yip! Yip! Whiskers twitching, a river otter galumphs along the mudbank and slips into the water. We linger, but, alas, the frisky otter doesn't reappear.

When the Colorado Division of Wildlife released these delightful creatures in the late 1980s, it didn't anticipate some of the driest years on record. And when engineers at McPhee Dam hoarded water, river flows dropped drastically. In a desperate search for enough water, the otters strayed as far as the Colorado River, seventy-five miles away. Such problems with insuring habitat speak eloquently for minimum stream flows. The choice is clear: if we want the thrill of seeing otters, we'll have to stop watering so much hay and so many golf courses.

River otters establish territories where they feed on fish and crayfish. But these superb hunters also share food with each other rather than fight over territory. Their ease of movement encourages them to visit one another, often just to romp. Like primates, they caress one another and cavort with liquid grace, often spinning as they swim. Aided by flaps that seal their noses and ears, they can stay submerged for several minutes at a time. Their sleek, elongated forms, with webbed feet and muscular tails, seem effortlessly propelled at up to six miles an hour. Otters turn everything into play.

Today, thankfully, playfulness and communication among otters, seals, dolphins, and whales garner our respect as signs of intelligence. This is a remarkable development because we humans have long defined intelligence in terms of attributes particular to ourselves—mainly language—which we've evolved more fully than other animals.

It's now late afternoon so we'll need to camp soon, well before sundown. Barbara calls for a landing. "Now you tell me!" I mutter as I heave hard for the bank. She teeters on the tube, then springs for the steep bank, skidding on mud. When she grabs a branch, it snaps in her fist and she plunges into the current, disappearing beneath the bow. By the time she surfaces, the current has dragged the boat from the bank. To compensate, I thrust hard into the oars, nearly submerging Barbara under the tube. She grasps for a willow stem and pulls herself out, spitting water and invectives.

"Tie us!" I yell with iron determination. She clambers up the steep, slippery incline, winding the rope around a willow. I pull hard against the current, managing to wedge the bow into dead branches that gouge flesh and drop spiders. Shivering, Barbara reels in the now bedraggled *Canyon Wren*. Before I can apologize, we wave in John and Lori, who land with more finesse.

Once moored we discuss landings, not feelings. Our first mistake was not to allow time for Barbara to reach the stern, a better place for the jumper and for the rower to pull the power strokes to hold the stern. Barbara wonders whether she'll bear the brunt of practice landings while we refine our moves.

Our campsite proves to be worth the trouble, however. Trees along the river shade us, and meadows offer open campsites. Wet clothes and gear hang on a pole that runs between the trees. As the sun sinks, John and I sit mesmerized by the reflections nearby: the bright buff of the sandstone blends into the light green from the opposite bank, the oceanic blue of the sky, and the tan of the river's swirls.

Our dining room takes shape beneath box elders. The menu includes *spaghetti al dente* with homemade *salsa marinara* washed down with robust *Chianti Classico*. For dessert? *Gelato* topped with dark chocolate and Milano cookies. Divine decadence in the wilds.

"Hey, who's the designated rower tomorrow?" chides John. With the river to haul food, rafters eat and drink well, sometimes too well, especially since rafting doesn't burn off calories the way backpacking does. In fact, despite the exercise of lifting, rowing, and sometimes swimming, it's easy to return with fresh flab.

We set up our tents near a patch of blue grama grass with

curved seed heads. Nearby John sees a three-foot snake using its head to shovel sand from its hole. Once it spots intruders, the snake vibrates its tail and flattens its head like a rattler. All of us except John recoil. "Hey guy, don't try to fool me—you're no rattler." In addition to burrowing, bull and gopher snakes exhibit mimicry. Since most predators hesitate before approaching a rattlesnake, bull snakes with mutations resembling rattlers enjoyed a greater chance to pass on their advantageous genes.

Yelling disturbs the quiet. Five kayakers who brandish beers, pitching the empties at each other, are followed by boatloads of yahoos looking like frat boys in hot tubs.

"Real men do kayaks," one hairy-chested guy blusters.

"Yeah, but they like rafts at meal time," John mutters under his breath.

When the "real men" on the river hear us chuckling, one spits his chew.

Why do some boaters, apparently oblivious to all this beauty, behave so irreverently in nature? Does the challenge of facing rapids arouse combative instincts? Maybe it's our culture's view of nature as a place to shed inhibitions that conditions such behavior. The biblical Fall characterizes the earth as morally and spiritually flawed, associates it with evil, and separates it from heaven. Nature is considered inanimate, literally devoid of spirit. Most Euro-Americans therefore find it hard to recognize nature's inherent sacredness.

From the viewpoint of ecopsychology, this estrangement is a serious problem. In *Nature and Madness* Paul Shepard contends that mistreatment of nature results from arrested moral and psychological development. Shepard, along with Robert Bly and others, laments the fact that our culture lacks rituals for bonding with the natural world or models of mature behavior in nature. Whereas Hopi infants are ritually introduced to the sun and exposed to revered elders who live in harmony with the outdoors, mainstream American kids are more apt to play with machines intended to tear up the landscape.

Granted, these kayakers have every right to release bottled emotions. But their constant splashing and shouting sound like echoes from the water park. Deeper wildness, on the other hand, unleashes the unconscious mind or taps primal impulses to connect the wildness inside with the wildness outside. This is unlikely to occur when approaching the river like a carnival ride.

Buoyed by the recent interest in environmental ethics, recreation specialists are asking which activities are uniquely suited to unspoiled natural areas, which degrade either the wilderness experience or the natural resource itself, and which might be better done somewhere else. Eco-ethics assumes a respect for wild nature. A floating kegger party belongs on Lake Powell, where it would not interfere with someone else's wilderness experience.

The first bats flutter in the twilight, diving to snare insects just above our heads. "Looks like we're drawing bugs," John quips. Lori swings wildly at the first mosquito, then waves in more bats. A chorus of night sounds whines and whispers. The four of us sit around our campfire, listening to it pop and hiss, sniffing the aroma of the dry juniper, bonded to warmly lit faces of friends. Suddenly a beaver splashes into the river. "Is that you, Barb?" John asks.

Toward dusk the main problem becomes finding the right path through the brush. I'm sitting on "the groover" when John and Lori stray off course, nearly walking into me. Should I flick on my flashlight, blowing my cover, or hang tight, hoping they'll find their way? They do, and I'm spared a beam in the face.

Soon I lie atop my bag listening to the shrill metallic cries of the nighthawks. These swept-wing predators are not hawks, of course, but members of the whippoorwill family that hunt insects. What incredible eyes they must possess to grab insects without the echolocation, or radar, that bats possess. Listening to the night sounds and watching for another shooting star, I feel intensely alive, at home in an unfathomable universe. During the night, a mystery bird utters three or four sharp whistles. Odd.

The next morning we enter the canyon wilderness. The walls rise faster than the sun, casting long shadows. Three distinct eco-zones suggest how crucial moisture is. Tall cane, privet, box elders, willows, and tamarisk beard the banks. Behind these, sage, rabbitbrush, and pea-green grasses cover the gentle slopes. Above them, dark-green pinyons and grayer-green junipers stud the rocky talus below the cliffs. Adding incandescent yellow to the scene, waxy prince's plumes candle the desert-varnished rock.

To find shade, we slip under a gnarled cottonwood that rustles like running water. A stronger gust rattles its deep-green leaves like rain splatting on rock. Dried by the desert air and sun, downy capsules on the boughs are breaking open. Hundreds of feet up, the fresh cotton fluffs stream like snow-flakes, luminous in the breeze. Violet-green swallows swoop down to the river, skimming its surface for bugs, then twitter high among the puffs. When these fliers finally rest, they perch in rows on a dead branch, turning their heads toward one another. As we push off, tassels on the cane sprinkle fine seeds on us; we're now the bearers of new life.

After a short hike, outstanding rock art gazes down on us

from a large overhang. Squiggly lines on one panel seem to reflect the trails left by ant lions in the dust nearby. Numerous inch-high figures appear here, possibly as a population count, plus a large figure with a hooded head and eerie strings hanging from its arms. Could such an imposing pictograph have possibly served the purpose of social control, much as art in many cultures is both religious and political? Today's Hopis, the descendants of the Anasazi, sometimes disclose that as adults they reexperience the fears they registered as children when they first saw kachina dancers. Whether a culture plans it or not, threats that "the boogie man will get you" can make people behave.

Afloat again, we drift through liquid space as cliffs flow upstream in a slow-motion revery. The canyon walls rise still higher to reveal glossy desert varnish that becomes irridescent in the dazzling light: black sometimes glares almost white. Rainstreaks stripe the blond undersides of a cliff before they fan out into paws of encrusted sand. The canyon walls swerve into alcoves eroded by ancient seeps or side drainages. These rank among the most spectacular red rock canyons in the West.

A rapid roars ahead, one that's not on the map. This is it: our first real test. Barbara calls out a stump and a giant boulder that pile up water. Downstream, nasty sleepers probably lurk beneath the turbulence. Just when we most need visibility, curves and drops make it more difficult. We watch for intermittent spray that spurts up high enough to see. Pulling down my hat, I backwater to buy time. Breaths come guardedly as we float the last fifty yards with my oars "chicken-stroking" last-minute corrections in course. The roar grows louder as the froth surges more visibly.

Then we rock and roll. The raft pitches on the smooth

pre-waves as the slap of a whitecap heralds bigger water. To blast the bow through the big waves, Barbara readies to punch the front tube with her shoulder. I watch the snag and the huge boulder. The sleepers loom in the corner of my eye.

Damn! Overcompensating to miss the snag leaves us heading right for that slimy boulder. I spin the boat from its forty-five-degree-off-the-snag position, then graze the boulder. We careen on, stern first and blind.

I swivel us around. We're riding the tongue of the rapid, its surface streaming like drawn glass. Its classic "V" leads right into the sleepers: a rock garden. Like a motorist who brakes on sheer ice, I foolishly try to backpaddle. Then I heave my torso frantically into a few strokes, hoping to cross the current.

Not enough. We're still heading for the rocks. Instinctively I drag one oar to straighten out just in time. Sunglasses drenched, Barbara looks around and yells, "Way to go!" We've navigated our first rapid, though we've got some bailing to do. Once the froth subsides, I kick back, sandals on the cooler. As John and Lori slice between the sleepers, John grins but Lori looks pale.

We park the boats and pick our way up to a flower garden a hundred feet above the river. Prickly pears still sport their waxy pink flowers, and dwarf daisies continue to bloom from their ground-hugging mats. Parched petals from globe mallows and firecracker penstemons mottle the bare ground where stunted lavender penstemons hang askew above the hot sand. Its oils activated by the heat, big sage perfumes the breeze.

Nearby a western collared lizard nine inches long flaunts its brilliant turquoise throat and belly. Then it dashes, yellow feet a blur. Collareds need this speed to catch other lizards. Though they're members of the iguana family, like most other lizards they've adapted to heat and dryness. Unlike mammals, these reptiles don't need to drink because they generate their own water from their food. Whereas mammals lose water to excretion, lizards eliminate not urea but uric acid, which requires less water.

A cicada case droops from a juniper trunk with only its form remaining: a blunt head, bent antennae, and gnarled legs. The insect itself has burst free, leaving the husk of its old life behind. Cicadas are still around, though, for their high-pitched whine thrums a pulse. In a bush nearby a small cicada flutters its wings, trying to break through the twigs. Finally it shimmies free to saw the air as it rises; others immediately join in. In one of nature's anomalies, male cicadas actually sing to attract other males. Generated by vibrating chambers in the males' abdomens, this swelling chorus attracts females that lack ears but use the walls of their bodies as sounding boards. To survive in the desert, cicadas cool themselves by sucking fluids from host plants and, to counter the intense heat, they can sweat off one third of their moisture each hour. Insects continue to astound us warm-blooded critters.

Back on the river, we enter the thirty-mile-long trench the Dolores has sliced into the Uncompahgre Plateau. Sheer walls tower above both banks. In most places the top layer of Navajo sandstone hundreds of feet thick outcrops as smooth bluffs. Below lie the thinner, softer, grayer Kayenta sandstone cliffs, plus benches littered with talus. Beneath them comes Wingate sandstone that forms the sheer, jointed, desert-varnished cliffs. Below the Wingate, the river is beginning to cut into the soft Chinle formation. Downstream the Dolores rasps deeper into the Chinle until it eventually reaches intractable black gneisses and granites. From slickrock to bedrock.

After more quiet water where box canyons indent the

bluffs, Bull Canyon appears as a real break in the wall, one we hope to hike. In contrast to backpacking, which often leaves hikers tired just when they arrive at the best places, rafting offers a way to reach remote areas while still fresh enough to experience them fully.

Like many side canyons, this one begins as a backwater. The trick is first to break through the brush along the river banks, then to get beyond the mud deposited at high water. While the brush poses no problem here, Bull Canyon is accessible only after a hundred-yard wade through muck. I slip in the slime but manage to thrust my old Konica skyward before splashdown. After we empty the gravel from our shoes, we enter a tight canyon. Blooming bushes thrive where rainwater occasionally sheets down the sheer walls.

After negotiating smooth steps in the sandstone, we inspect a brimming bowl. Schools of beady whirligig beetles zigzag erratically to generate waves that serve as echolocators. When disturbed, the whirligigs dive. Water striders jerk about, sometimes stacked three high. This pool also supports short-legged water striders—really vellids or "skeeters"—that look like miniatures of the larger variety.

A monarch butterfly sails by, cinnamon orange against the overarching blue. Monarchs are no flutter-bys, however, for each winter they fly to Mexico, where they face destruction of their forest habitat. On their long migrations they struggle over mountaintops, generally flying over rather than around obstacles. Though typically solitary, monarchs congregate by the thousands and navigate by magnetic fields that allow them to fly after dark. By late summer their flits from flower to flower will cease to be random as they feed on their southward voyage.

The monarch is often confused with its look-alike, the

viceroy butterfly. Since the monarch feeds on milkweed containing alkaloids, birds vomit soon after eating one and associate the discomfort with the black-and-orange insect. By mimicking the monarch, the viceroy benefits from the bird's misidentification. Despite their keen eyesight, few birds have evolved the ability to distinguish these two insects by sight. However, some jays have learned how to eat monarchs by cleverly pulling off the legs and wings, which contain most of the alkaloids.

Above us, some ordinary-looking white daisies droop from a hanging garden. On closer inspection, though, these are rare kachina daisies, or fleabane. The Dolores Canyon is one of the few places where kachina daisies grow in Colorado. They're also uncommon in Utah, where botanist Stanley Welch discovered them near Kachina Bridge in Natural Bridges National Monument.

Of the many reasons why a plant becomes rare, the most common is specialization. Kachina daisies demand specific conditions, such as a particular mineral content in a seep, that do not occur in many places. Sometimes, too, a plant's pollinator becomes uncommon or extinct, reducing its ability to reproduce. Other times genetic drift or hybridization produces new competitors.

A final reason for rarity occurs when the plant is stranded by changes in climate. Like many disappearing species on the plateau, kachina daisies are probably leftovers from the Ice Age when cooler and moister conditions prevailed. Now such relict plants can live only in scattered refuges. If global warming changes the climate faster than plants can adapt, extinctions accelerate. It is natural for some species to disappear, but the increasing rate of human-caused extinctions is alarming.

The huge question, then, is how to slow the current epidemic of extinctions. Among many approaches, Pulitzer Prize-winning biologist E. O. Wilson urges nothing less than changes in attitude: "The more closely we identify ourselves with the rest of life," he suggests, "the more quickly we will be able to discover the sources of human sensibility and acquire the knowledge on which an enduring ethic, a sense of preferred direction, can be built." In short, if we can rediscover our heritage, we can not only save thousands of species whose lives hang in the balance but also save ourselves. But we are *homo erectus*, the stiff-necked species that doesn't stoop to sniff the earth or kiss the ground.

Since it's late, we decide to camp here at Bull Canyon. A swarm of gnats dances just above reflections of wavering cane. John jokes about getting extra animal protein as he refills his wine cup. But soon these blood-suckers become ravenous. The females are the most voracious because they need blood before they can lay eggs. We daub on Muskol and Cutter with little effect. The gnats move in. John tries an old trick, coating his skin with cooking oil. When he starts to look like flypaper, the cure proves worse than the affliction. As soon as small flies join the gnats, Lori ducks into the tent. Her retreat compounds her woes when the tent collapses around her: the sagged nylon and its human pole make a forlorn sight. The rest of us swallow chuckles.

At dusk bats begin to skim the water. Suddenly the bugs are gone. Amazing. Bugs typically fly when their metabolisms heat up and when they won't become dehydrated. But the temperature and humidity didn't change in a moment, so what happened? Could these gnats and flies sense the ultrasounds that the bats navigate by? After all, houseflies possess hairs that pick up the vibrations from movements in the air, such as those of a fly swatter; these gnats may also hear ultrasonic squeaks.

As John and Barbara finish the dishes by flashlight, a hearty fellow appears, his light blue eyes sparkling under his cap. When he asks our names and then repeats them, I invite him to pull up a log. A raft guide from Colorado, Mark is a raconteur par excellence. He's regaled many a boater around the campfire. As his stories of river lore wane with the embers, he swigs the last of his wine and returns to his camp. Mark lives simply so he can do what he loves most: running rivers.

The throbbing roar of Bull Rapid nags John's and Lori's sleep. Earlier John had bushwhacked downstream for a look, but he didn't like what he saw: a rocky bank precludes either lining or portaging their canoe around the rapid. This morning Lori is relieved when Mark offers to paddle in her place. With Mark at the helm, the little canoe deftly dodges the rocks. Barbara and I pull directly out into the current to run the gauntlet, no problem.

Clear skies prevail as gentle breezes silver the sandbar willows. Between gusts two bumblebees dart out from the bank, circle Barbara in her yellow life jacket, and then beeline back, disappointed. Heralded by their characteristic "tree-eeet," a flock of sparrow-sized sandpipers scoots along the mud flats on spindly legs. As we approach, two or three of these birds (also known as "peeps") sound the alarm with a series of whistles. As they take flight, their tawny wings beat rapid but shallow strokes just above the river. They fly in tight formation, each banking as they land abruptly on a sandbar across the way. That mystery bird jaws again, camouflaged in the heavy brush.

A common merganser, which is really a pied duck, leads

her procession of ducklings. Her head sports the merganser's characteristic "wings," but her mate's greenish-black head lacks such feathering. Mom and brood dive in unison, with the young surfacing first. Unlike other ducks, mergansers use teeth-like protrusions on their bills to grab fish during their long dives.

When another raft glides by, Lori calls out, "Are there any rough rapids ahead?" A tall boatwoman shoots back, "Just below Spring Canyon—stay left but don't get thrown into the wall." We stop first to hike this canyon; we'll scout the rapid later.

About a mile up the canyon a drama of life and death unfolds in a shady pool. The usual characters—toads, tadpoles, boatman beetles, water striders, and whirligig beetles are acting out their roles, but the plot takes a turn when an ant bites me and I brush it into the pool. Before the water striders, instantly sensing surface vibrations, can do much more than pick at the struggling ant, a new player enters from above: a long-legged spider drops from its conical web, seizes the ant, and returns to its lair, upside-down. I later identify this as a long-jawed orb weaver, a common spider that clusters its legs for concealment and leaves a hole in its web for quick passage.

On our way back ants appear in much greater numbers. Like brown strings, three lines of ants descend from top to bottom of a fifteen-foot sandstone cliff. Lined-up antenna to abdomen, these amazing insects disappear into tunnels in the rock and emerge fully a foot later. We know that ants secrete their own pheromone scent trails, but how did they discover these passages? Since few ants carry cargo, what's the purpose of this mass migration?

Each partial to a different route, the four of us head through the foliage to scout Spring Rapid. Barbara's field glasses locate a narrow channel between the left wall and a cluster of rocks. She steps back from the viewpoint, unsettled by the ominous rumble and turbulence.

As soon as we've launched, I pull hard strokes to carry us across the current. We hover for one last look, then plunge down the chute. Mindful of the warning to avoid the wall, I overcompensate to the inside and snag on a submerged boulder. Splashing does not free us. Barbara and I try heaving with our bodies. No go. I yank an oar from its lock and, thrusting it down hard, free our bow. We spin into the current out of control. Feet spread for balance, I jam the oar back into its lock. As we emerge, I notice that John has waded into the current to help, which means he is risking his bad knee. After watching our travails, John and Lori line their canoe around this one.

Later we beach at the broad mouth of Coyote Wash, one of the highlights of this trip. When tired hikers skirt our campsite, they rave about the scenery and rage about the deerflies. Cutting across a large meadow of needle grass, I hike around a classic perched meander, a place where the river cut a channel that it abandoned eons ago. Blown sand and fallen boulders now fill the river's old course to illustrate the carrying power of the river, which is constantly clearing its present channel. As John and Barbara cook, Lori rubs Benadryl on her bites. I page through my field guides, scribbling notes and sputtering about the mystery birds that won't sing out their genus and species.

The next morning we rise before either the sun or the deerflies can hammer us too hard. A stream we camped near has risen during the night, probably because it wasn't losing much of its flow to evaporation. To my amazement the creek is rolling mud turds, globs of blue clay that settled out at high water. Given its typical depth of two inches, the load this creek carries is astounding. Upstream, tongues of sand advance rapidly enough to make a photographer backpeddle.

As we wait for John and Barbara, Lori and I observe this tiny creek. "Look at that standing wave," I remark, reading its inch-deep current as a river runner. Lori sees the sparkling rills as fish scales or a wash board. "This isn't a river," she chides, disposed to any subject other than rapids. Maybe I'm becoming fixated on reading rapids, much like a climber obsessed with crevices in rock walls. In *A River Runs Through It*, Norman Maclean observes how fishermen see rivers mainly in terms of fishing. Every way of relating to nature soon becomes a way of seeing.

Coyote Wash is a broad-bottomed canyon that meanders for eight miles down from the La Sal range east of Moab, Utah. After getting hit so hard by bugs yesterday, John, Lori, and Barbara wear long pants today. The dreaded deerflies move in, and before long it looks as though we've disturbed a hornets' nest; hungry dark clouds stalk each of us. Lori and Barbara swat madly and break into a trot; John and I keep striding while we swing our shirts like horses' tails. Soon we're all swatting away with our sandals.

Damn! Deerflies gnaw John's back and my legs. We run for shelter under an overhang. There the marauders subside, though they linger ferociously in the sun, not far away. These are strange insects. They're smart enough to bite in the tender places but stupid enough to stay put and get swatted. Actually, the issue is one of adaptation, for deerflies probably evolved feeding on mammals that didn't have hands for fly swatters. We slog on, gauzy and baggy, looking like beekeepers caught without smoke machines.

While the group snacks, I explore alone. The canyon runs

about a hundred feet wide, its stream braiding into rivulets. For some reason, probably because of flash floods, little grows on the broad canyon floor except for a lone hackberry tree that quivers in the rippled air. As the floor undulates its way into an incline, the sinuous slickrock rises for several hundred feet. Bright blue pinyon jays squawk in the junipers far up sheer sandstone walls. Orange butterflies dance above a pool as a tiger swallowtail glides just overhead, its backlit yellow wings glowing like a Chinese lantern. The deerflies are long gone.

Fully comparable to the gorgeous Escalante canyons, Coyote Wash is not only one of the most scenic hikes in the Southwest but one of the most sensory. Because its walls slope gradually here, its floor receives enough sunlight to nourish soft grasses. With its flat, sandy bottom, it's a sensational place to move like an ancient Greek athlete, without clothes, just for the sheer kinesthetic joy. Bare feet experience the textures of sand and water, bare body savors the temperatures of air, sweat, and sun.

On the way back, John hails me from a sandy bank. As we recline in the cool sand, John pokes a stick into a cone in the sand. He's baiting an ant lion to flick the sand that makes its quarry skid into the pit. Ant lions, really the tick-like larvae of insects resembling the damselfly, are common in desert country that otherwise supports relatively few insects.

As John recalls his boyhood misbehavior toward animals, including a lurid banquet of birds and frogs, he concedes that he owes wildlife some good deeds to offset his earlier misdemeanors. Dredging my own memories, I confess to shooting frogs with my slingshot. Perhaps children, especially boys, treat animals cruelly because they lack power and find that they can exert it over animals. Boys can also be compassionate, though. A boyhood friend and I used to arrive at school late because we'd been rescuing earthworms stranded on sidewalks.

Across the way a furry brown animal bounds from a puddle. We look at each other: "What is *that*?" This critter stops, runs again, and finally rolls in the loose sand. After it shakes, it rears up like a prairie dog. Then it scurries along a ledge and rests in the shade. By the time I'm able to locate my binocs, it's disappeared into a blowhole in the sandstone where it turns in circles, looking out each time. After watching it for a while, we conclude that this is a desert-dwelling rock squirrel. Like their arboreal counterparts, these rodents sport salt-and-pepper coats. Rock squirrels are well adapted to a desert; if their food supply dwindles because of drought, they may begin hibernating as early as August.

Something sprinkles us with sand. As we look up, a fat lizard spins in the sky. To our amazement, two whiptail lizards are dangling in midair, tails quivering. Shameless voyeurs, we watch them mate and wonder, since many whiptails reproduce asexually, if these are both females with one in the role of the male. Suddenly the pair drops to the rocks six feet below. Just one lizard reappears, tail whipping wildly. We've witnessed high-wire romance with dire consequences.

After generous gulps from canteens laced with electrolytes, we stride back into the afternoon sun. Salt replacements prevent dehydration and, more importantly, make the canteen water taste potable enough to down the prescribed gallon per day. After a few hundred yards, we soak our hats, shirts, and pants to cool ourselves rapidly in the dry air. For years I've assumed, no doubt self-indulgently, that less clothing meant better cooling. But in brilliant sunshine like this, maximizing ventilation, evaporation, and interception of solar radiation is the best way to go.

Less than a river mile below Coyote Wash, we marvel at a dramatic saddle in the canyon wall. A whole canyon system seems to exist back there, complete with dazzling white cliffs. Then I realize that this must be Mule Shoe Bend where the river takes two miles to travel the two hundred yards across this isthmus. In a half hour or so we'll loop around to the other side of the saddle; over geologic time the river will eventually erode through to shorten its course.

A few miles farther, we discover another great place to play.

A straight slot, easily seventy-five yards long, marks the mouth of La Sal Creek. In contrast to the Dolores, which runs cold and cloudy, this tributary flows warm and clear over gravel swept clean of silt. Its current sweeps me along, tummy up, much as the Platte River carried Loren Eiseley. After the rapids, which usually require resistance, it's good to float with the flow. As we recline under the willows, a sharp "whoit" comes from a thicket, but no bird shows itself—only a great blue heron, head poised to peer through its own reflection, patiently fishing the shallows.

As soon as we launch, a rapid takes us by surprise. About two hundred yards down from La Sal Creek, the river dives around a bend. Distracted by beavers swimming in a placid backwater, we hadn't noticed that the river takes a turn. Our ears are suddenly filled with the sound of rushing water. I even stand on the rowing seat, but it's still impossible to see what lurks below. My throat convulses.

Playing the percentages, we follow the curved tongue. This is standard technique, but here the tongue heads straight for treacherous rocks. I backwater hard, butt braced on the back of the seat, buying time to read the channel. "Right,

right!" Barbara shouts. Pushing hard into a stroke, I skirt the first two rocks. A wave slaps her in the face; she wipes her eyes to locate the next hazard. We careen off one boulder, spin and carom off another like a pinball. Pools of bubbles whirl and fizz as we emerge from the turbulence. Broken sticks swirl in eddies.

The Dolores has tested us and we've passed our preliminaries. We've tasted the adrenalin rush and crave more. We eddy out just below the rapid, wondering how John and Lori will handle it. After absorbing the fury of the afternoon sun for a while, we spot them lining their canoe along the far bank. Paddling decisive strokes on the same side, they shoot right across the current and head for shore.

On the beach the four of us fan out to reconnoiter where little grows except speckle-pod locoweed. Firm sandbanks invite us to camp just above the river's edge, where small waves lap rhythmically. To prepare for a fire, John and I gather aromatic pieces of dead juniper. Despite all the driftwood, there's no human trash, not even a broken styrofoam cup. This is a pristine stretch of river.

We pitch camp, glad for moments of shelter from the blowing sand. Twilight lingers while John and I sip wine, mesmerized by the river's lapping, eyes riveted to the great wall far above that incandesces in the amber light. This Entrada sandstone layer, smooth but cupped, peachy in regular daylight, now tints more toward salmon. This canyon country is, as W. L. Rusho eloquently observes, "a land where earth tones are daily enflamed by the rising sun, colors change constantly as shadows creep about, diminish, and lengthen throughout the hours." After the cantaloupe tones fade, we sit silently in the light of the Milky Way. Enchanted by it all, we almost forget to kindle our fire.

Illumined from inside, our tent now outglows the embers that still exude the delicious scent of juniper. When Barbara blows out the candle, I'm left with the river, where Buddhist thoughts swirl in my mind. We're riding on a river, I muse, whose speed, size, and course lie beyond our control. We steer around life's hazards, but submerged rocks lurk, unforseen, and the current sweeps us in new directions, some of them unwelcome, others mortal. When we say "go with the flow," we blithely assume that the river will carry us where we want to go; we seldom anticipate that a seemingly perfect tongue through a rapid will lead to a big drop or a black hole.

As my stomach begins to roil, I slip into our tent and snuggle against Barbara. Only yards away the Dolores, the river of sorrows, laps at the sandbank.

The early sunrise flushes us from our tents. After lingering over coffee, we get on the river early. As we float lazily, silence reigns until eventually we begin to hear a five-hundred-ton drilling rig clanking. A service road gashes the river bank. Here the Bureau of Reclamation is drilling a sixteen-thousand-foot-deep well as part of its Colorado River salinity control project. The Dolores not only hauls a heavy load of silt, it also carries ten thousand tons of salt a year from Disappointment Creek alone. These desalinization projects have come about because Colorado River water is over-allocated, and its diminished flows are often too salty for agriculture.

On this last slow stretch, the brush comes alive with sharp whistles, discordant scoldings, and the sharp "whoits" that we've been hearing so much. Show yourselves, you teases! Then, from a perch on a dead branch, an arresting beauty with a sulphur breast, an olive-brown back, and a long, graceful tail performs not only one but all three of the calls. The three mystery birds coalesce into one: the yellow-breasted chat. Enthralled with this large wood warbler, we nearly float right by the takeout.

Mark, the raconteur from upstream, strides up as I bemoan the gouges on the bottom of the *Canyon Wren*. Ready to engage his bronzed, muscular torso, he offers to help tear down our rig. "How come the river is dropping so fast?" John asks him. As a river guide, Mark's ready to protest "messing with rivers." He spews some salty invectives to condemn malpractices at the McPhee Dam. In the late 1980s, he contends, the dam operators choked the Dolores down to a mere trickle, which stressed the otters and killed the trout. What Mark doesn't mention, though, is that during some summers before the dam was built the Dolores used to dry up entirely.

"The Bureau of 'Wreck-lamation' spent half a billion dollars on that dam. Now that it's done, farmers can't afford the water. Has the dam helped the local economy? No. And they don't just capture spring runoff, they rob you of peak flows."

Mark has a point. The Dolores is diminished, her ability to support wildlife like otters compromised when the engineers spin the spigots. Rare plants, including the kachina daisy and the Dolores rushpink, still survive in this gorge and its side canyons, but many native plant communities also require both flooding and adequate stream flows. The Bureau of Land Management has (bless their souls) recommended the surrounding area for wilderness designation, but this stretch of river also needs protection under an official Wild and Scenic designation. More than ever before we need to preserve wild places. When we lose them, we lose a part of ourselves.

We jam everything in John and Lori's Bronco. Barbara and I squeeze into the back, sandwiched between the door and life preservers. I pray that a dumpster is not far, for the back windows don't roll down and our garbage is reeking. Since the highway back to Slickrock gives no hint of the glories that lie in the gorge, my mind's eye flies high above to trace a beige ribbon, richly trimmed in green, that loops through a deep redrock trough. I wish we were there.

Interlude

Bats and Bugs

To humans, bats seem silent except for the whuff of their wings. To gnats and other insects, however, the "silent" darkness screams warnings. Most bats hunt by echolocation, a radar based on emitting cries and picking up the echoes. Using sounds beyond human hearing—in fact, often well into the ultrasonic spectrum—bats can determine their quarry's speed and direction. They even factor in the well known Doppler effect, the tendency of sound from a moving source to increase or decrease in pitch. Based on frequency modulation known as FM, bats' cries are perfectly adapted to locating small prey at short distances.

With their aerial agility and their sophisticated radar, bats would seem to experience no more difficulty tracking a bug than a chameleon would have grabbing an ant. But during the fifty million years that bats have hunted, insects have evolved some very sophisticated defenses. When bats come closer than ten feet, many insects adopt a variety of acrobatic maneuvers including erratic flight, rapid turns, powered dives, and spirals. Flying crickets, for example, react to a bat's sonar by swerving or even diving into a free fall. Certain moths mimic the bats' squeaks to jam their radar. Tiger beetles fly toward the ground and land; katydids stop flying entirely. One prey species may display several different moves to keep bats from anticipating a single evasive maneuver.

Equally amazing is the timing of insects' responses. Crickets, for instance, change their wingbeat rate in only sixty milliseconds. Other insects detect the telltale ultrasound but react only if the danger comes too close. The fact that insect evolution has selected for this variety of moves has forced bats to act more intelligently, learning from experience to anticipate the evasive movements of their prey. As we look further and further into the infinitely complex evolutionary process, we find that we humans are far from the only species to evolve greater and greater intelligence.

7.

Moonlit Faces on the Wall

Snow drifting into silent ruins, icy winds neighing in the rimrock, a great horned owl cruising, silent as death.
—Rob Schultheis

Like mangroves in a lagoon, rounded junipers spot the uplands of Cedar Mesa where blue-gray sagebrush floods the broad drainages. As it takes flight, a flock of mountain bluebirds flashes azure feathers. Purple vetch and dwarf red paintbrush are already blooming, and bright spring grass rings a dormant anthill. Lovely as it is, this mesa doesn't hint at the spectacular canyons that incise it, hundreds of feet below.

There, where snow clings to the north-facing walls of Bullet Canyon, ice still rings the shaded pools. Bleached beeplant stalks bristle along the stream. Except for a single fly basking on a protected boulder, the insects remain inactive here. An Anasazi watchtower reminds John and me that we're entering the Grand Gulch canyon system, one of the glories of southern Utah and one of the world's great archaeological areas. In fact, we're hiking where Richard Weatherill, the early cowboy digger, built a stock trail before the turn of the century.

It's mid-March, so we're descending into spring. After hibernation in a crevice, a mourning cloak butterfly warms itself on a rock, velvet wings quivering, legs stretching back to life. It feebly flutters into the sunshine where it rests, slowly fanning its chocolate-brown wings until the warmth gets its juices flowing. Soon it whuffs by my face and lights on a yellow pussy willow. When another mourning cloak dances by, the pair pirouette skyward into the deep-blue sky.

John and I stride down a giant chute where the canyon bottom rounds into slickrock walls, shelves, and bowls while the horizon narrows. These colorful walls are Cedar Mesa sandstone about 250 million years old that runs from four to twelve hundred feet thick. As early geologist Herbert E. Gregory observes, classic Southwest sandstones differ: "crossbedding in the Cedar Mesa seems neither so prominent nor so uniform as that in the Coconino and resembles little the great curves of the Navajo."

Though this bedrock weathers into slabs that make fine footholds, with such heavy packs we zigzag and lean inward toward the walls. We lower ourselves into a bowl, toes groping for footholds. While John wonders how he'll climb back up these drops, I ponder what this narrow trench is like after a cloudburst.

As though to welcome back admirers, the canyon regales us with the melodious call of a rufous-sided towhee with a black head, white markings, and a brick breast. Throat bulging, he sings a melodious "drink your teeeee" (popularly rendered as "towhee") to attract a female. In the meantime his song signals "no trespassing" to other males that might enter his territory. As he forages through last year's leaves, he hops backward noisily, his red eyes alert for anything that crawls.

Beneath the flume stand Fremont cottonwoods scarred by flash floods. These cottonwoods commemorate John C. Fremont, the early explorer and botanist who camped in their welcome shade. Trees that grow here absorb seasonal beatings from flash floods, which usually occur in August. With their deeply furrowed bark to minimize the wounds that can lead to infestation by insects, cottonwoods can take these hard knocks. One twisted trunk stretches across the sand like a fallen log, then rises into a tree whose knobby twigs end in swollen buds. After it got knocked down, this scarred survivor got back up, alive and ready to grow.

Its deep greens set against the buff sand, an inconspicuous chimaya blooms nearby. A member of the carrot family, the chimaya sprouts clusters of golden flowers above its fleecy fronds. This and other early spring plants spread mats of dark foliage just above the ground to pick up reflected sunlight and lingering heat. Above us, on a talus slope, a round-leaf buffaloberry unfolds its pale-yellow buds. Farther up, other buffaloberry bushes silver the steep slopes. A plant unique to the Colorado Plateau, the buffaloberry derives its unusual silvery quality from tiny stalked hairs that, under magnification, resemble the dried tops of Queen Anne's lace.

Since chimaya, buffaloberry, bladderpod, wallflower, and other early bloomers sport yellow flowers, they indicate that insects which pollinate spring flowers are attracted to yellow. Many of these are flies, which are essential early spring pollinators. Hummingbirds, which are drawn to red blooms, don't arrive until bugle-shaped flowers such as scarlet gilia and penstemon bloom in May or June.

Beneath a slab of sandstone crawls a bumblebee in its brown pupal case. This dormant bee feels cool but soon warms to life in my hand, where it feebly stirs its crude claw-like legs. Once back in the cold sand, however, it attempts to dig back in. There's no rushing growth; this bumblebee needs more time and more warmth to mature. Later on it will become a cinnamon-colored adult fully an inch long, ready for flowers of all colors. As I carefully replace the slab over the insect, I consider that regard for living beings increases the esteem we feel for ourselves.

Further downcanyon, deeper into spring, a pool is full of life. As I peer in, a wolf spider streaks for cover. Water striders shoot across the pool, their feet denting the surface film and casting oval shadows on the sand and slime below. The Anasazi sometimes used images of striders to mark the center of petroglyphs, apparently because of their cross shape and affinity for water. The smaller beady-black whirligig beetles are more frenetic. In response to predators below and above the surface, they have evolved two pair of eyes, a nearly unique adaptation in the vast world of insects. Spiders, striders, and whirligig beetles all hibernate, so when spring comes they must feed and breed before these pools dry up, as they often do.

As the canyon widens, our eyes scan the walls for Perfect Kiva Ruin, abandoned centuries ago by the Anasazi and reconstructed recently by the Bureau of Land Management. A wall once enclosed this cliff dwelling, possibly to retain turkeys, dogs, or children. Below the kiva lies a midden heap resembling the caked feces of a pack-rat burrow. Since midden heaps served as both trash dumps and burial grounds, they've often attracted diggers, legal and otherwise.

This is called Perfect Kiva Ruin because it was discovered with the kiva roof intact, which is remarkable because cattle have trampled so many sites. A ladder allows us to descend into the nether world, a place where Anasazi males communed with ghostly kachinas—and with the sacred earth. Shelves and storage holes large enough for a hand are indented into the wall. Rocks surround the fire pit, and the hole in its center represents the *sipapu*, the navel of the earth through which all life emerged. Blackened adobe masonry still coats the walls, probably because the ancient ones often burned their buildings when they abandoned them. After minutes of revery my mind drifts toward wonder.

Emerging from this subterranean sanctuary makes the canyon above look overexposed, surreal: I squint and rub my eyes as though waking into a world revealed. Did the Anasazi hold their religious ceremonies underground not only to engage the earth but also to emerge into a spiritualized realm of radiant light?

Above the kiva only one residential room remains, but it is well preserved, with its stone-and-mud masonry unbroken. Inside a T-shaped door, possibly designed to afford better entry for someone bearing a load, a shaft of light shoots through a smoke vent—an unusual feature. Outside, John locates other relics: grooved boulders used to sharpen sticks, miniature corn cobs, bleached squash stems, twisted willow ties used to fasten branches for mud-and-wattle construction, and pieces of yucca fiber still wrapped around the turkey feathers that once lined blankets. To touch what the Anasazi made is to connect with them, to wonder how many fellow humans did this blanket warm? At a time when vandals have scrawled obscenities into priceless petroglyphs, artifacts such as these speak well of previous visitors.

In fact, hikers are now helping to catch and prosecute backcountry vandals. In 1992 hikers observed a professor from Ricks College in Idaho who looked on as his sons scrawled their names on precious pictographs in Canyonlands National Park. Other hikers reported this to a ranger who apprehended father and sons. Later, according to the *Federal Archaeology Report*, the three faced felony charges and paid stiff fines. But crimes against rock art are hardly the only threat, for illegal pot hunting has been a weekend hobby in southeast Utah for a long time. In recent years it's declined, partly because hikers now photograph thieves, but it still goes on.

Under the overhang a "yellow man" pictograph inscribes the buff sandstone and mud balls remain right where they stuck, over seven centuries ago. Perhaps mischievous children threw these balls when the masonry mud was still wet. Or perhaps, like the one-inch sandstone balls nearby that were used for hunting or for games, adults threw them in some more organized context. I heave a similar-sized rock as high as I can, watching it strike far below the highest mud balls. Did someone propel these balls with a stick? It's well known that some Southwestern tribes used atlatls, or throwing sticks, to shoot projectiles with greater force.

After this immersion in prehistory, we camp in the next alcove. As I sprawl on the bedrock, deep breaths settle me into a bowl where my body flows like putty, filling the basin. These cemented grains of sand have stayed intact for over 200 million years, and most will remain firm far beyond my lifetime. In relation to this rock, my life is as inconsequential as the grain of sand that breaks loose in a spring thaw. In *Wind in the Rock* Ann Zwinger speaks to a similar experience with time: "I could sit here for eons and watch as these sandstone walls crumble, grain by grain, and fall to floor this dry wash. . . . The rock is ephemeral, the wind, eternal." In this sandstone, however, I also see the ancient waves and winds frozen in time, with each ripple affirming a world where sand grains either blow in the winds or trickle in an hour glass.

Nearby Jailhouse Ruin takes its misleading name from sticks in windows probably intended to keep people out, not in. Like smoke holes, windows were relatively rare in Anasazi architecture. An inaccessible upper tier of rooms, plus the wall protecting them, were built well: their masonry is still solid. A long slit in this wall apparently shielded archers or served as a chute to roll rocks onto potential intruders. Farther along this ledge stands a "guardhouse" that looks up and down the canyon.

While defensive sites do not necessarily imply hostilities, they do suggest that these Anasazi definitely paid attention to defense. Possibly because these were outposts—located "on point" relative to Grand Gulch—they required more protection. The close cultural similarities between, say, the Mesa Verde and Grand Gulch Anasazi may mislead us into concluding that there were few intratribal conflicts. To assume this would commit the mistake of thinking that because Mayan cities resembled one another closely, they did not fight each other. Security was clearly a factor in Anasazi life, though it's not exactly clear why. One clue might lie with the clay figurines that anthropologist Sally Cole found here during stabilization. These exhibit a distinctly Fremont Indian style that could suggest forced incursions into Anasazi territory.

As the sun halos a lonely cloud with gold, it also gilds the rock ramparts. As we toast near our fire, shadows creep up the canyon walls. Doves whir, seeming to send a warning. A few other birds cheep, then everything falls quiet except for the wind that whistles through the ruins.

Since one learns to filter out a lot in today's world, it's exhilarating to open all the senses. For minutes on end all is still. Suddenly a mouse rustles the leaves behind us. Immediately an owl hoots hoarsely, whooooo . . . whooooo, echoing from all directions at once. The mouse continues rustling, oblivious as it scratches for food. This mouse lives dangerously, for the uncanny hearing of long-eared owls enables them to hone in on mice, even those that scamper under the snow.

Many cultures have believed that owls embody souls in torment or other sinister tendencies. While the Greeks revered owls as symbols of victory and wisdom, they also regarded them as prophets of doom. Both the Greeks and Romans, believing that bad repels bad, tried to ward off evil with images of owls. Farmers in some cultures still hang a dead owl on a barn to scare off other birds. Owls are impressive to both eye and ear, especially at night, but as a result they've tended to become associated with a lot of fears and superstitions.

Moonlight now floods the rocks across the way. This is not the usual bluish moonlight since it glows slightly orange, tinted by the lingering twilight. Here, more perceptibly than in Grand Canyon, moon shadows retreat down the canyon wall. There, because the walls often remain a great distance away, it's difficult to see moon shadows moving. Here, however, the walls rise right behind our campsite so we can watch planetary motion as rhythmic as the tides. At day's end, sun shadows rise; after dark, moon shadows ebb.

As our fire dies down, we slide into our bags, shivering,

to wait for the moonrise above the rim. We place bets on when it will be. Five minutes, then ten. Still no moon. At last the sky silvers, silhouetting scattered trees on the mesa top. Finally a marble mound bulges from the slickrock: the full moon beams into our rapt faces.

In the brilliant moonlight the circles the Anasazi painted stare at us like watchful eyes. Some archaeologists believe that these circles are sun and moon symbols used in solstice rituals, while others note that circles are prominent icons in cultures where the reverent whirl or dance in a circle. Native American writer Paula Gunn Allen indicates that her mother taught her how "'Life is a circle, and everything has a place in it.' That's how I met the sacred loop."

Nearby the moonlight animates three "dancing man" figures. We're not the first visitors to sense strange energies near these ruins. The Navajo have long believed that the spirits of the Anasazi haunt these sites and have even tried to exorcise them by defacing petroglyphs. Earl Shumway, a notorious local pothunter, also tells strange tales. In an interview summarized by Robert S. McPherson in *Sacred Land, Sacred View*, Shumway recounts digging an Anasazi grave where a rattlesnake bit him: "He dreamed of 'three long knives leaping out of the ground and stabbing him in the heart,' which he associated with the body in the grave. He had uncovered an albino medicine man who had a 'little pouch full of arrowheads and pouches of different smokes, herbs and pipes.'" Later a wall collapsed on Shumway. He returned a second time only to break his ankle, and a third time on crutches only to encounter a nest of giant red ants. Yet even this was not enough to deter Shumway.

Our breath now steams in the cold silver light. During the night I rejoice that we have down bags instead of turkey-feather blankets. After a long night, the morning sun sparkles on the frost. The slush clogging our canteens needs melting before we can brew coffee. Like campers around a fire—no, like an amoeba engulfing its food—my body encircles the stove to absorb every calorie of heat. John inhales the steam from his coffee as he takes breakfast in his bag.

Freed from our back packs, today we'll explore downcanyon. Hiking with only day packs and canteens, we set out for lower Grand Gulch. After dead ending in a thicket of willows and tamarisk, we retrace our route to the main trail. From higher ground, this tangle looks inviting: its pastel colors— pinks, oranges, browns, greens—blur together like an impressionist watercolor. Treading over the tracks of mice, pack rats, and other nocturnal critters, I feel like the first human in an untrodden desert garden.

More pictographs decorate the walls along our route. One roughly grooved panel presents two dominant figures with broad, perfectly flat shoulders, slightly tapered trapezoidal shields or bodies, and drooping hands and feet. Except for the absence of heads, these are typical of Basketmaker Period "little men," or anthropomorphs. These pictographs may represent gods. Polly Schaafsma, an authority on Utah rock art, suggests that these imposing figures "not only had ceremonial import but . . . were probably representations either of supernatural beings themselves or of shamans." The Anasazi may have assumed these images to contain the spiritual force of the beings represented. The handprints that appear around the supernatural figures may have identified people who offered prayers to the divine beings.

Mystery certainly enhances the fascination of rock art. To interpret it, we try to approximate the beliefs of ancient cultures drastically different from ours. While this is probably impossible for non-Indians, the Pueblo peoples of New Mexico and the Hopi of Arizona, both descended from the Anasazi, do offer some clues. Even these tribes, however, suffer from a cultural disconnection, and our limited understanding of rock art archetypes may contribute to our sense of having lost our way.

Nor are attempts to understand the Anasazi mere academic exercises. The "ancient alien ones" survived in the arid Southwest for centuries and evolved complex responses to this environment; their experience can teach us things. In many areas, for instance, they thrived until their wood cutting led to erosion that jeopardized their canyon-bottom farming areas or until their numbers had outgrown their fuel and food supplies. The Anasazi did not just disappear: erosion and overpopulation, exacerbated by the long drought of 1276-99, apparently coupled with disease and incursions by enemies to hasten their demise.

After reaching the confluence, we enter Grand Gulch. Here, where the canyon walls open to let in more sunshine, spring is more strongly under way. As we ply the brush, some catkins and pussy willows are yellow, others lavender or lime green. Black carpenter bees and small flies luxuriate in these fuzzy spring blooms. In one alcove the dark gray branches of large hackberry trees, their bark ruggedly ridged, feather subtly into a cliff in the shade. Boldly colorful as this canyon country is, it also exhibits shades of gray that would have thrilled master photographer Ansel Adams, the master of such subtle nuances.

A hairy woodpecker flies rhythmically, much like a kingfisher. It utters a "peek" as it spreads its wings to land on a cottonwood, then it hammers out several volleys, stopping to look around after each burst. Drumming serves the same function for woodpeckers that singing does for most other birds: it attracts the opposite sex. We're probably within this fellow's territory, which woodpeckers often define between favorite perches. So we move on.

Enticed by a side canyon, we tread softly up a sandbank of soft spring grass. Reveling in its softness, I wonder how long this grass can last once the sun begins to blast this sandy soil; grasses in arid regions go to seed quickly and remain dormant through long droughts. In a thicket of box elders, withered leaves from last year slow the upsurge of the new growth. As I sit and ponder, orange-and-black box elder beetles that have hibernated under the dried leaves are crawling on my boot.

Life abounds beneath this leaf litter. As famed biologist E. O. Wilson suggests in *Biophilia*, a clod is not mere dirt. Beyond the innumerable micro-organisms, soil houses dozens of insects, mites, spiders, nematodes, and earthworms. The genetic data contained in one handful of humus would over-flow a large library. Multiplied infinitely by all the handfuls of earth everywhere, all this information comprises the basis of ever-continuing evolution.

Like it or not, we're connected to all this life. We share breath with plants, imbibe the same water as animals, and carry much of the same genetic makeup. Equally famous biologist Lewis Thomas remarks that "For all our elegance and eloquence as a species, for all our massive frontal lobes, for all our music, we have not progressed all that far from our microbial forebears. They are still with us. Or, to put it another way, we are part of them."

Out of our insecurities or vain wishes to feel superior, we fear both cultural differences and the implacable otherness of nature. If we consider nature's own purposes, which ulti-mately include our own, it becomes apparent that diversity is crucial. Genetic differences allow each species to adapt, survive, and continue the flow of evolutionary change. From the viewpoint of ongoing life, of regulating the earth's atmosphere and temperature, grasses are probably as important as humans. Poet Walt Whitman understood that sprouting leaves of grass imply rebirth, miracles, and interconnections. When I get up to leave, I find that I've left my imprint on the spring grass, and it has left wrinkles on me.

With hardly a sound, a magnificent red-tailed hawk swoops down the side canyon, utters high-pitched cries, and does a circle-eight before soaring away. Redtails are buteo hawks, raptors with heavy bodies and long, wide, deeply notched wings. With a wingspread often reaching three feet, thick wings with five "finger" feathers, and scalloped gray and cinnamon plumage, these are truly impressive birds. This hawk is screaming at us, the latest intruders in its hunting territory.

In this side canyon we walk barefoot through new grass and sunbathe fully—two rites of spring. The sun's rays throb on my closed eyelids, spawning catsup-colored dots. A curious bee fans my face, momentarily blowing away the warmth from the sun. As John rests, I slip away to explore barefoot, amazed at the varied temperatures of the moist sand. I hone in on the hum that animates a huge old cottonwood literally crawling with honeybees. Carried through another winter by their honey, these harbingers of spring pursue their comings and goings through the vibrant air. Twigs quiver as the bees suck. This male cottonwood is dropping sticky bud bracts to unfold maroon, pollen-rich catkins that resemble kernels of Indian corn.

Nearby this great tree of life a brook heralds an enticing side canyon that snakes away from Grand Gulch. This drainage soon narrows down as smooth walls rise hundreds of feet overhead, curling in and out of drainages. After meandering about a mile, it ends in a pool below a drip. I feel like the only human alive.

A small ruin up the cliff begs for exploration. But how to get there? There are no footholds in the rock, no makeshift ladders left by other explorers. Ignoring my weak ankle, I look for shelves to approach the ruin by sliding along the rock face, toes on a narrow ledge. A juniper tree, its loose and stringy bark offering little assurance, enables me to reach the first ledge. As I stand up slowly, fingers groping for grips, I wonder whether this ledge is wide enough to sidestep my way toward the ruin. This looks possible, at least to where the cliff curves out of sight. The rock face sandpapers my knees when the cliff gets steeper.

As my hands perspire, they pick up loose sand that hardly helps me grip the rock. One at a time, ever so slowly, I wipe them in my hair. After thirty-five feet of toeing along the ledge, face against the cliff, I can't go on; it's just too sheer. O.K. I could return along the ledge, but if I'm going to have to jump anyway, why not jump from here and spare myself having to cling on the way back?

No, that would be a stupid thing to do. This is the first serious hike since my accident. Last spring I got exuberant, pranced on some boulders, and tried to bounce off one that wobbled, spun, and dropped me down. My badly crushed ankle kept me hobbling all summer. If I mess up here, I might do permanent damage. And John might not locate me for days. This is no place to play desert bighorn and jump off a cliff.

How to turn around? Since my canteen won't fit between me and the rock, I toss it down, swivel very carefully, and spider down with hands and feet gripping. My good ankle will have to absorb the impact. Skidding slows me down, but I still hit the sand hard and stumble into a yucca blade that punctures my calf. The blood drips down my leg but soon coagulates. Luckily my weak ankle feels no worse.

Before I can compose myself, however, a block of sandstone drops with a hollow thud. Broken slabs grate down the talus slope, sounding like crushed bones. Dust drifts into the grainy beams of sunlight. Strange. Why would this happen now? It's nowhere near as hot nor as cold as these rocks get, so contraction should not be the cause. I might suspect the spirits of the Anasazi whose home I'd tried to visit but my mind doesn't accept supernatural causes—at least not in daylight.

Rock falls seem fast and infrequent, but they actually represent the cumulative effects of unseen minute changes. "So slowly, oh, so slowly have the great changes been brought about," concluded naturalist John Burroughs: "One summer day . . . a section of stone wall opposite me . . . suddenly fell down. . . . It was the sudden summing up of a half century or more of atomic changes in the wall. A grain or two yielded to the pressure of long years, and gravity did the rest." And so it goes with human change as well; after a long, dark winter of discontent, of changes too small to notice, warmth returns and people burst into bloom.

John and I return to Bullet Canyon, enthralled by the glowing colors. Past the confluence, a wall of spires like those in the Needles District of Canyonlands thrusts toward the sky. Two rounded rocks resembling geodes cap the top layer of creamy sandstone. Back at our campsite, the painted eyes watch us from a distance.

More fuzzy contrails stripe the sky. Why are there so many jet trails in this area? When the military continues to commandeer still more airspace in the West, it often channels commercial flights down narrow flyways. In fact, so much of Nevada's air space is restricted to military aircraft—Nellis Air Force Base is larger than any National Park—that civilian planes often must zigzag across the Great Basin. Given the military superiority it already exhibits, why does this country play expensive, consumptive, and destructive war games? Regardless of which machines intrude on our experiences in nature, their disturbance isn't just momentary. They remind us of the security-hungry society we'd hoped to forget, diminishing our awakening consciousness of life's interconnectedness, of wilderness still beyond industrialism.

Tonight will feature neither a sunset nor a moonrise, for the glare from overcast skies grays the rocks. As purplish rings encircle the moon, we retreat to our bags. John reads a Tony Hillerman mystery, while I scribble notes in my journal, switching from elbow to elbow until they both ache.

After a night of damp cold, gray dawn seeps into our tent. Soon an insistent pitter-patter sounds the alarm. I pull my bag over my head in hopes that the rain will go away; instead it begins to freeze and bows in our tent. We'll have to face the weather before we get snowed in. We pull on heavy clothes and eat quickly, huddled in our sagging shelter. Then we plunge into the storm. Wind and rain now chasten my exposed back as I hurriedly roll up the nylon, sand and sticks and all.

We ready ourselves for a long hike out. From beneath these rain hoods we won't take in the scenery; we'll see our boots today. Once underway I watch the stones underfoot, their colors intensified by the rain. For the first mile or so I feel deprived of sights beyond the drops on my visor, beyond the streaks of the rain. I feel alone, encapsulated in plastic, although John hikes only ten feet away. With visibility reduced, we lose the trail often and mumble irritably. No familiar landmarks indicate that we're making headway; we can't even tell if we're in the right canyon. My weak ankle wobbles, raising doubts throughout a moody, meditative morning that descends into self-pity.

As the rain becomes sleet, my mind turns inward. First I try to become the pain in my ankle, then I focus on the pleasures of movement. Boots and jeans already soaked, I splash through the cold puddles like a drenched schoolboy. Later on shivers of joy accompany a sense of cozy self-containment as the sleet splats on my raincoat. Without mental guidance from above, the physical me climbs over slippery rock, stays balanced, and ignores the pain. Mind and body approach unity, though mind stays ready to attend to the body's calls for concentration. The present prevails as the cadences of my steps and my breaths coalesce, three footfalls per inhalation, three per exhalation. Sharp cold air in, soft warm air out. Now beyond self-pity, I generate a glow affirming that, yes, I'm happy to hike here, rain or shine, fully alive.

The slickrock, especially in the chute, has become very slick indeed. The runoff spouts from the rimrocks, splattering into plunge basins like a pitcher pours watermelon juice into a bowl. As John picks his way up the rock steps, I wonder what his rainy day has felt like. We trudge on, legs shaky from the cold and the strain, my ankle now rolling. The Anasazi guard tower appears through the sleet. "Not much farther, we're almost there," I pant. Finally there's the break in the wall. I climb out on all fours so as not to slip on the slush. At last, the mesa rim.

We hug in celebration—raincoats, wet packs, cold hands, and all. I pull back my hood and take snow in the face to see all around. Winter's back big time. Clusters of winged crystals frost the junipers whose bare branches protest against the snowy sky and bend under the weight, their shriveled berries never having appeared so gray. Cedar Mesa doesn't look like this every day.

Although my VW van whines and coughs and chugs, it finally does start. The heater, however, doesn't kick in until forty miles later, just as we arrive in Blanding. We're looking for a cheap motel but we can't pass up food. Any well-heated restaurant will do. With mud-caked jeans and matted hair, we look like drowned desert rats.

This place is very Utah, a decidedly white-bread state where copies of Dan Valentine's *American Essays* often sit beside the napkin racks. Without so much as a glance at the menu, we order the large country-fried steaks. Before we can hit the salad bar, though, we find that this is where the guys bring their gals for a pre-prom dinner. While we tear into our grub, boys fiddle with their cufflinks and girls adjust their corsages. We try to keep our muddy jeans under the table, aware that we probably detract from the romantic atmosphere in other ways as well.

As John and I chow down, my mind wanders westward. Back in Bullet Canyon the butterflies have slipped into a crack, the bees are fanning their wings against the cold, and those cottonwood catkins are holding on against the wind. Beneath the faces on the wall, the spring snow is falling gently on early yellow flowers and on a hunched hawk, now too wet to extend a feather. As the owl hoots, mice freeze and rabbits shiver. Silences follow before the owl flaps into the flake-filled air. Often we expect spring to arrive with no wintry "relapses," but springtime in the Southwest comes in ebbs and flows.

Interlude

Birds Raining from the Sky

Eared grebes in their fresh gray and white winter plumage were flying at night. On their migration from the Great Salt Lake to Baja California, these red-eyed diving birds began thudding down through the fog and snow onto roofs and windshields. Apparently attracted by asphalt that glistened like water, the survivors proved helpless because grebes can't take flight from land. Benefactors in the Utah town of Beaver boxed them up and released them on lakes, but half of the grebes died soon afterward. Nor is this the first such incident. Storms forced down mass-migrating birds in December 1928 as well.

Fortunately, nature provides for such losses. To allow some of their chicks to survive, grebes lay so many eggs that they oversupply predators. Evolution could not, however, anticipate the human disturbances that have caused eared grebes to decline. About one million, over half of the world population, fly south from the Great Salt Lake each fall. But winter habitat has become a problem. Because of over-appropriation of the Colorado River, water levels have dropped drastically. Concentrated in these dwindling backwaters, toxic chemicals from agricultural runoff have affected birds by the hundreds of thousands. According to the Fish and Wildlife Service, such toxins don't kill birds outright but instead weaken their immune systems, which leaves them vulnerable and eventually reduces their ability to breed. Hatchlings often exhibit grotesque deformities. Did poison-induced diseases weaken these grebes, making them more desperate for shelter?

While the Great Basin west of the Colorado Plateau still ranks among the continent's great flyways, bird populations are dropping for other species as well. As Terry Tempest Williams wonders in *Refuge*, how can we protect threatened waterfowl? The Western Hemisphere Shorebird Reserve Network has identified Bear River Migratory Bird Refuge and Fish Springs National Wildlife Refuge as crucial habitats, which they certainly are. While these wetlands do provide stopovers for water birds, they aren't entirely safe havens; one of them experiences destructive fluctuations in water levels and the other abuts a bombing range and a proving ground for chemical weapons. Because synthetic reservoirs store water but often don't offer essential food-rich habitats for birds, preserving the salty, temporary lakes of the Great Basin is crucial for many birds' survival. In rainy years, at least, invertebrates such as tadpole shrimp proliferate in these "playas." When the playas are full, a feast fuels the birds for their strenuous journeys.

Especially in the arid Southwest, it's essential to halt the destruction of the few remaining wetlands. Here more than elsewhere a missing link in the chain of splashdown spots can finish off a flock of hungry, tired, possibly weakened fliers. Given the excessive thirst of sprawling communities such as St. George and Las Vegas, saving wetlands will be difficult. Most Americans feel compassion when they see birds in trouble, but few understand that by the time birds lie stunned on the pavement, it's often too late.

8.

Dinosaur in March

I saw a distant river by moonlight, making no noise, yet
flowing, as by day, still to the sea like melted silver
reflecting the moonlight. . . . By it the heavens are related
to the earth, undistinguishable from a sky beneath you.
—Henry David Thoreau

Since nobody runs the Green River through Dinosaur National Monument in March, we'll experience it much as the early explorers did. Our friends Matt and Barb will lead in their raft; Barbara and I will follow in ours as we face our first big water.

Despite its dangers, this stretch tells a long story of river running. In 1825 General William Henry Ashley and six mountain men confronted the worst rapids on the entire Green River. In search of new grounds to trap beaver, they set out in bull boats (buffalo hides stretched over pole frames) adapted from Plains Indians. The Canyon of Lodore impressed Ashley powerfully: "As we passed along between these massive walls, which in a great degree excluded from us the rays of heaven . . . I was forcibly struck with the gloom which spread over the countenances of my men." With little choice but to continue, the Ashley party became the first to navigate the rough water we now face.

The second party encountered more problems. On his trek to California in 1849, Captain W. L. Manly did not relish wintering in Salt Lake City. Operating on the widely held misconception that the Green River flowed west to the Pacific, Manly surmised that he'd rather float than walk or ride. So he and six other Forty-Niners dug an abandoned ferryboat from a riverbank near Green River, Wyoming, and pushed off for California.

Theirs was indeed a "voyage of discovery." Indians tried to hail the party, but Manly paddled right by them. Below Flaming Gorge the crew tried to fend off rocks with poles, but the powerful current wedged their ferryboat between boulders. To survive, the men had to carve dugout canoes. Just above what Major John Wesley Powell later named Disaster Falls, Manly found a boat with a note indicating that a previous party had given up here to hike out of the sheer-walled canyon. "Disconcerted" but undaunted, Manly and his men persisted. After laborious portages around boulder fields and one mishap in which Manly nearly drowned, they rode through the rapids of Dinosaur. Downstream, in the Uinta Basin, they encountered a band of Utes who dispelled their delusions of floating to California. Sadder but wiser, Manly abandoned the river and headed for the place he set out to avoid, Salt Lake City.

From here in Browns Park, which the Manly party crossed in their makeshift canoes, the Green River seems to enter a mountain, seemingly an odd place for a river to cut a channel. Geologic processes explain this paradox. About twenty-five million years ago the river swung eastward, around the ancestral Uinta Mountains, to join the North Platte rather than the Colorado River. Sometime between twelve and fifteen million years ago, however, the eastern end of the Uintas sank nearly four thousand feet, opening up a potential channel. As the Rockies rose in what is now southern Wyoming, they backed the flow into what is now Browns Park, forcing the Green to grind its way through the sunken end of the Uintas. Then, while these red rocks slowly rose again, the river began to dig its half-mile-deep trench. Today dark-red cliffs finely powdered with snow impressively portal the Gates of Lodore. The Red Creek quartzite towering two thousand feet above us is about a billion and a half years old—newly risen but much older and harder than the sandstones of the Southwest.

It's quiet on the water. There's only the slight whoosh of oars stroking the river like a cook smoothing a sauce with a wooden spoon. Tiny whirlpools spin off the oars, leaving bubbles. Ordinarily our road-and-city consciousness—full of worries about work, money, or whatever—persists for a day or two as we enter the natural world, but in only a few hours the river leads us toward another consciousness, one that's more focused and present.

"Ahonk, ahonk." Male Canada geese emerge from the shade, their honks keeping time with their wingbeats. Perhaps these Canadas are heading back to Browns Hole Wildlife Refuge, a prime nesting ground for them as well as for egrets, herons, coots, killdeers, and raptors.

Just ahead, at the mouth of Winnie's Grotto, Major John Wesley Powell found a boat abandoned by trappers. Downstream lies what Powell, after losing one of his boats, named Disaster Falls. His crewman Jack Sumner describes their problems there: "The No Name . . . struck a rock and swung into the waves sideways and instantly swamped. Her crew held to her while she drifted down with the speed of the wind; [she] went perhaps 200 yards, when she struck another rock that stove her bow in." Mindful of this mishap, Powell's second expedition took two hard days to line and portage around Disaster Falls.

Just above the falls, broken rock fans into the channel from both banks where a massive fault enables two side canyons to enter and dump debris. One side canyon will usually produce a rapid; two spell trouble, especially when the channel slopes sharply and sheer walls also drop boulders into the channel. Barbara and I wonder if we really want to try this. We see our own images in each other's sunglasses, two munchkins dwarfed by a mountain of sheer rock.

Following Matt's expert lead, we dodge sharp rocks through Upper Disaster Falls. Matt and Barb pull over to applaud our run through this "rock garden," but suddenly my oar strikes a rock and bolts from its lock, almost bludgeoning my skull. The oar disappears into the turbulence. As he bounds over boulders, Matt shouts directions that are lost in the guttural roar.

This is serious. We can't stop. We're plunging into Lower Disaster Falls with only one oar. As I track the loose oar in the froth, I grope for the spare. Jagged rocks and nasty holes rivet my attention. A wave crashes from behind.

"Watch for that damn oar!" I sputter from the beleaguered captain's seat. Our boat spins out of control, careening off one rock before grazing another. We're taking on more and more water and becoming less and less manageable.

I fumble for the spare once again, but the boat lurches too much for me to unsnap its clasp. Finally the oar comes loose and fits into the oarlock. But it doesn't give much control in the maelstrom—its sleeve has loosened so it slips up and down in the lock. Our new equipment is failing its tests.

Froth again engulfs the lost oar. Our boat bounces off one rock, first dropping into the hole and then spinning out. I backwater hard against the current, but we skid into another boulder, bow scraping. We scramble to the high tube so the river doesn't wrap the boat around a rock. If we're stuck here long, we'll never see that oar again. We heave with all our might. More grinding. We slam the tubes again as though they were blocking dummies. Finally the current sweeps us off the boulder.

Eureka! Our lost oar bobs in an eddy. Barbara tosses a line to snag the oar; it's all beaten up but it's back on board. Only now do I realize that my heart is racing and I'm out of breath. After navigating Lower Disaster in fine form, Matt and Barb chide us for not securing our oar. I'm relieved that they don't know about the dysfunctional spare.

We land just above a big one: Triplett Rapids. Tomorrow we'll get wet early—just how wet is the question. No human signs appear on the beach, only the droppings of geese. Where they exist at all, paths reveal deer prints and droppings. At this time of year, at least, this area seems untrammeled. Its pristine quality also results from Park Service restrictions on river runners. Here in Dinosaur the issue isn't limitations on numbers, which most everyone supports; it's preferential treatment: permits issued for commercial parties definitely outnumber those for private trips, and outfitters in Colorado and Utah net millions of dollars a year.

Near our camp a Canada goose struts and pumps his head. When this gander hisses and lowers his head to charge, I retreat, though not fast enough. He hisses again and feints with his beak, then he strikes with his wing. My shin stings. Not only have I stumbled into a male defending a nest, but I've been faked out by a bird. Canadas are instinctually smart enough, however, not to continue such aggressiveness. Later on, after the goslings are launched, the gander should relax.

These geese display intelligent patterns in flight as well. When the lead goose tires, it rotates back in the formation while another flies point. Flying in a V enables geese to create an upward air current reducing fatigue and allowing for greater flight range than if each bird flew alone. The geese in the rear honk to let the leaders know that they're keeping up. Humans may dismiss all this as mere instinct, but it's intelligent behavior nonetheless—and it shows a concern for the common good that's becoming rare in the human realm.

To reduce anxiety, Barbara and I toss our rolled-up safety rope back and forth like a football. After massive cliffs swallow the tired sun, the sky turns steel gray and distant stars flicker in the twilight. Colder air settles; we huddle around the fire telling jokes that divert my mind from the unsettling experience at Disaster Falls. After everyone has burrowed, my eyes fixate on the vibrant red-and-gray coals. Like the sap that oozes from the driftwood, my confidence seeps into the sand. I gulp back my own fear. For the first time in my life, I want off a river. Soon I snuggle next to Barbara, wondering if she senses that I'm not asleep. The realization that I've entered the realm of wild creatures who don't know the meaning of worry helps to quiet the chattering monkey inside.

In the morning we devour steaming eggs and sausage,

shifting from leg to leg until Matt bellows "oinkers away." As he and I scout Triplett Rapids, he decides that we'll enter left of the snags, let the current sweep us toward the wall, and then stroke hard to avert a giant boulder.

"Long rapids are tough because either you don't see the hazards at the bottom or, if you're spun off course, you're not in position to dodge them. Even when you can follow your plan, you've often got to cross the channel against big waves." Not what I need to hear.

Matt points to standing curlers which are deceptive because the water of the waves races on while their shape remains stationary. After the current drops down a fall, it often bounces off the bottom and springs up again as a whitecap. As far as I can see, it's impossible to determine any regular sequence to these upsurges. They illustrate chaos theory in very relevant ways. These curlers turn anticipation into anxiety.

Matt and Barb shoot right through, making it look easy. We follow our plan, but, despite the strongest strokes I can muster, we almost graze the dreaded boulder. As we shoot by, part of the river gushes between this rock and the wall to create a powerful current, one we couldn't see from above.

While we stop for lunch, Matt illustrates how to cross the current within a rapid. He chuckles knowingly when I ask him about holding back a move—much like a pilot leaves some play in the stick for a sudden crosswind. "You bet, and here's how." As we kneel, he sketches river, rocks, and raft on the sand and suggests that one way is to spin the raft, pulling with the current.

"It's risky because you can't see much over your shoulder, but if you're goin' to make headway across the current, you'll have to sacrifice some visibility." I've been discovering that the greater the boat's angle to the current, the greater the effect of my strokes. When Matt extends this principle to include strokes that enlist the power of the current, I add the "downstream ferry" to my repertoire.

About a century ago modern river running techniques evolved on this very stretch. Whereas Major Powell had tried to run rivers in ruddered, round-bottomed boats that plied the water bow-first, in the 1890s trapper and prospector Nathaniel Galloway tried smaller, flat-bottomed skiffs with blunt sterns. Dispensing with the rudder and rowing alone in a cockpit, Galloway began to enter rapids stern first, but at an angle. This allows the rower both to face hazards and to apply powerful strokes against or across the current. The result is the widely used Galloway Method.

Although talking technique helps restore my confidence, Hell's Half Mile roars just downstream where our main task is to avoid boulders. Higher flows reduce this technical problem but also increase the speed of the current. Before the dam shackled it, the Green thundered down this millrace like a tidal bore. One of Powell's "ten who dared" describes the fate of one boat:

> The stream was so swift that it caused great rolling waves in the center, of a kind I have never seen anywhere else. The boys were not skillful enough to navigate this stream, and the suction drew them to the center, where the great waves rolled them over and over, bottom side up and every way.

While it's intellectually interesting to know that a large river rolls boulders from its channel, right now it's not reassuring to learn that river waves, unlike those in open water, tend to suck a boat toward them, right into the madness.

Hell's Half Mile is a blur of speed and spray. We watch

how Matt sets up, estimating that he's poised three boat-widths from a marker rock. After Matt's boat rides the crests and troughs, we plunge into the chute. Smack! Our bow thrusts skyward, and Barbara seems airborne right above me. Another wave comes piling in like a whale's tail slapping a longboat. Nothing exists but waves and rocks. Barbara heaves her weight against the front tube to drive us through a rooster tail wave; she's a natural at punching tubes. From here on the trick will be to decide early enough which way to dodge rocks and then to dig deep strokes so as not to fan air on the crests of waves. Timing is crucial: miss a stroke and the river can gobble you up. This is all river on the river's terms.

We emerge soaked but right side up, glad that the toughest rapids are now behind us. Matt pulls off at Ripping Brook in the deepest part of the Canyon of Lodore, 3,400 feet below the highest rim. Up there, atop the older, harder Uinta Mountain quartzite, lies a bed of pink-and-tan Lodore sandstone hundreds of feet thick. An exceptional sandstone, it contains fossils of trilobites, brachiopods, and other early marine invertebrates. Though most sandstones reveal few fossils, there are grand exceptions—most notably the sandstone strata in the Morrison formation here in the Monument. In this world-class fossil quarry, paleontologists have unearthed a massive boneyard where a Jurassic-era river dropped dinosaur bodies onto a sandbar.

The four of us picnic under a giant ponderosa pine, its bleached trunk stained russet-and-amber. Though gusts sweep its upper branches, sounding like ocean surf, they scarcely ruffle a feather here at river level. Water ouzels—or American dippers—dive deep underwater in search of fish or insect larvae. By beating their wings with great strength, ouzels can hunt on river bottoms where currents are too strong for a human to stand. And by exploiting scales that close their nostrils, plus third eyelids that seal against sediments, they can even hunt in rapids. One by one the dippers surface and bob on a rock, possibly to signal each other above the roar that would drown out an auditory call.

Douglas firs creep down the canyon wall and tuck themselves into the ledges. By sinking their roots into crevices, these lovely trees can grow from the rock. Just beneath their dense, compact crowns hang the characteristic cones with three-forked tongues, and farther down the sweeping boughs droop long, slender twigs adorned with a spiral of needles. As these trailing pennants wave frond-like in the breeze, they reveal reddish-brown, scaly bark that contrasts with their bluish-green foliage. Since this, the Rocky Mountain variety, is more resistant to drought than the coastal Douglas fir, its lighter color represents an adaptation to conserve moisture by absorbing less heat.

When upcanyon winds blow harder, we make little headway on the river. In fact, we row past a rock only to find ourselves blown back into it. "Dodging the same damn rock twice is . . . ," I mutter. My mouth hangs open, mid-word.

Three bighorn sheep stare at us from the willows, a ram and two ewes. The full-curl ram snorts and rises, nostrils flaring, his horn three inches thick at the base and his yellow eyes bulging so he can see in many directions. We drift still closer, but the bighorns don't run off. Finally all three spring up to a higher point to watch us; they can see sharply for up to five miles. Bighorns reintroduced into the Canyon of Lodore in the 1950s are now well established, though they remain a rare sight in most of the West.

Prior to the coming of Euro-Americans, a remarkable variety of wildlife inhabited this area. No less than five ungu-

lates—deer, elk, bison, bighorn sheep, and moose—suggest
the original complexity of this ecosystem. But with the intro-
duction of livestock during the nineteenth century, indigenous predators also fell victim to the white man's deadly traps
and guns. The Department of Agriculture's Animal Damage
Control (ADC) agency, for instance, trails a bloody history.
According to Lynn Jacobs in *The Waste of the West*, "Since
1915, the federal government alone has reported killing
roughly a million bobcats and lynx, primarily for stockmen."
Large packs of wolves still roamed Browns Park until the turn
of the century, though professional exterminators had nearly
wiped them out by 1906. Grizzlies, of course, are gone
entirely. To extend Thoreau's image, these missing species
represent pages ripped from nature's book, one we read with
sorrow. To know natural history is to feel what we've lost.

Today we might expect that federal agencies would move
predators from areas where they conflict with human interests
and then release them where they could help restore damaged
ecosystems. However, in many places the Fed still wages war
on wildlife: in 1988 alone it poisoned, trapped, or gunned
down 76,000 coyotes, 300 bears, and 200 mountain lions.
Lacing carcasses with poison is still common practice even
though it destroys non-targeted animals, including our national bird. In 1993, in a sly move that would have amused
George Orwell, the ADC changed its name to "Animal
Services." One wonders what services the program delivers
besides death.

Beneath the two-thousand-foot Tiger Cliff the headwind
builds in Echo Park. Where manganese oxide streaks this
smooth expanse of Weber sandstone, its black-and-blond
stripes gleam in the brilliant sunshine. From an alcove under
this stupendous wall, Matt's bellows echo through an enclo-

sure of several square miles. Downstream looms Steamboat Rock, a huge trapezoidal monolith dominating the confluence of the Yampa with the Green. The flood-happy Yampa remains the last free-flowing major tributary in the Colorado River basin.

Like the obsession with predators, our fondness for dams reflects a control-and-subdue mentality. Just below here, at the head of Whirlpool Canyon, the Bureau of Reclamation proposed a seven-hundred-foot dam that would have backed up both the Green and the Yampa rivers. In one of the great conservation battles of the century, David Brower of the Sierra Club led the fight to stop this project, aided by writers such as Wallace Stegner. *National Geographic* sponsored river trips and ran articles challenging the dam-builders' contention that Echo Park was "inaccessible" without the dam.

Most importantly, in 1938 a proclamation had enlarged Dinosaur National Monument to include the canyons of the Yampa and the Green. This enlargement, as historian Roy Webb observes, enabled "the inviolability of the national parks" to become a central issue in the long controversy. Conservationists finally prevailed in Dinosaur, but they also made a tragic compromise: they consented to the even more destructive dam that flooded two hundred miles of Glen Canyon. As a result, one of the gems of the Southwest now lies submerged beneath Lake Powell, lost forever.

We push off for Jones Hole campground, seven miles downstream. Whirlpool Canyon offers some of the most colorful geology in the West, a gorge where many layers, each painted in pastel earth tones, are convoluted in astounding ways. The current picks up, amplified by colder, grayer water from the Yampa, but the upcanyon gusts prevail. Barbara hunches down to reduce wind resistance. Leaning bodily into strokes, I groan. To ignore the pain, I concentrate first on how my oars enter the water, then on how deep they penetrate before they surface again. Calves against the thwart, I lean into deep, slow strokes with my whole being. This is one of the most spectacular stretches of river anywhere, but I'm laboring too hard to look at it.

While Barbara takes the oars, I lean back on wobbly elbows to gaze at the incredibly uplifted, fractured, and bent sedimentary rock. On one side of Whirlpool Canyon, the Mitten Park fault rises; on the other, the Island Park fault stands different layers on end to fashion vertical stripes. Here the familiar Navajo sandstone of the Colorado Plateau is bent and contorted yet its colors look almost as surreal as the Painted Desert.

Where mosses choke a discolored creek, Jones Hole seems unpromising at first. The human impact is apparent but tonight, amazingly enough, we have three whole campgrounds to ourselves. As our fire crackles and roars, primitive urges make me yearn to blow the flute or spin like a dervish. Instead we sit, all too civilized, and cast long shadows.

The cold and the early darkness, however, do motivate us to do something with spiritual possibilities. Soon the fire dies down to expose the stream-rounded rocks glowing red hot among the coals. Using a canvas, six oars, and rocks from the creek, together we fashion a "tipi" sweat lodge. The hot rocks hiss with steam. We undress quickly, crawl through the flap that serves as a door, and sit around the rocks, heads hanging loose. Before the pearls of sweat can melt, the heat drives us into the chill, cleansed and enlivened. Afterward, though, I feel a vague disappointment at not having had a more transcendent experience. But my spirits have been lifted, which is perhaps all I can expect given the limits of my cultural background.

At the trail register the next morning, Barbara checks entries left by other visitors. She chuckles at what someone has scrawled: "Name: 'E. Abbey'; Destination: 'Glen Canyon Dam'; Nature of Visit: 'Pyrotechnical Research'; Remarks: 'Hayduke Lives.'" Much-needed comic relief. After the magnificent palette of colors we've seen on the river, the cliffs along Jones Hole Creek seem drab under these leaden skies. Behind us, the canyon wall across the river remains in the shade, still blotched with snow. All seems wintry except for a few songbirds chirping in the brush. Where are the signs of spring?

Before long, though, the world greens and Jones Hole becomes a plant paradise. Easter daisies find their niche on sunny, rocky benches where their early blooms spring from a mat of leaves. Yellow nuttall violets, their leaves mottled a greenish gray, bloom on the sunny side of a rock. Nourished by freshets, watercress and monkeyflowers trim the creek. Their roots protruding from the bank, river birches send up dozens of finger-sized shoots that snare leaves, mosses, and other debris. These stems bend with the flow, for birches are pliable. Unlike the seeds of most birches, however, those of this river variety sprout on the fresh silt deposits as the spring torrents recede. Evolution selects for the germination time and place best suited to the habitat.

The sun blazes through as a brown trout streaks into a bubble-filled pool. Fresh from hibernation, a mourning cloak butterfly suns its cream-and-chocolate wings.

On the other bank it's "Just-spring," or "mudluscious," as e. e. cummings called this time of year. Protected from freezing along the stream, mosses glisten a vibrant spring green. Rivulets course everywhere to flatten grasses like windblown hair. Frog eggs bubble in raccoon prints, and box elder fruits sprout their chartreuse seed leaves. Flattened this-tles are spreading, and round-leaved burdocks are pushing aside the mud. In fact, great clusters of burrs grasp my cuffs. Around my boots, thousands of black-and-orange box elder beetles crawl sluggishly on the matted leaves, ready to feast on the new growth.

The pictographs in Jones Hole are equally sensational. Here pictographs left by the Fremont Indians seem un-touched by the centuries. Toward the middle of the panel, in pale yellow, stands an unusual figure that resembles a New Mexico cross sprouting feathers or horns. The headdresses or horns on several little men suggest animal antlers. Many archaeologists believe that headdresses or horns signify shamen or dieties.

Further along appear three more humanoids—probably gods—with shield-like torsos that are again topped with feathered headdresses. While trapezoidal torsos and earrings characterize Fremont culture, the hair bundles evoke the Anasazi style. Both Anasazi *kokopelli*, or hunchback flute players, and their *kolowisi*, or horned serpents, appear throughout the Uinta Basin in the heart of Fremont territory. Wandering Anasazi traders may have ventured north, bring-ing artistic motifs.

"God, a bighorn," Barbara whispers. I fumble for the binoculars, and soon we count nine. Two rams dominate the top of a huge boulder, while one nudges the other's rump with his nose. Their massively based but gracefully spiraled horns indicate that they're at least ten years old. These older males appear to guard the herd, yet they remain the farthest from danger. A ewe leads two lambs toward us, followed by the herd. The lead ewe stretches to see over a rock, sniffing the air and snorting softly. She stares at us, slips behind the rock, then steps still closer, watching through an amber eye with a slit down the center. This ewe and her young are the point animals and the old rams are posturing, not protecting.

The ewe half bellows, half bleats. She appears above us, then stops leading the herd. She cries out again. Apparently we are crouching near the chute where the herd descends to the stream. While many bighorns don't drink every day, pregnant or lactating ewes need water daily. We move down-canyon to watch. The ewe descends a bit but no farther. It's time to leave the bighorns' canyon home.

As usual, supper is hearty fare from the dutch oven, a down-home meat-and-potatoes meal that evokes traditional Western ways and does not interfere with our purpose of experiencing the river and the earth. Around his small fire Matt recounts memorable questions posed by people on his raft trips: "Do we take out where we put in?" and "What time of year do deer become elk?" After we all howl, Barb adds another classic she heard at Mesa Verde: "It's a great park, but why'd they build it so far from the highway?" Well, at least these visitors asked questions. Too many don't even do that.

Come sunrise, Matt is already clanking his fire-blackened oven. Today we'll run about fourteen miles of mild whitewater and see some truly spectacular riverscapes. In the day and a half we've spent at Jones Hole, the river has come up a foot, which means it'll give us a faster ride. We all bundle up to face the frigid spray. No sooner are we in the current than we're suddenly heading into a two-hundred-foot-wide rapid I'd have liked to scout. Though we skirt the boulders, a few four-foot brown curlers do spill over the tube. Teeth chattering as she bails, Barbara complains that we're hitting all the biggest waves. I'm secretly glad for any whitewater that, by counteracting the sharp upcanyon wind, reduces the interminable rowing. I've already faced my trial by wind and water.

In Island Park, where the walls recede and the Green finds space to meander, welcome sunbeams warm the air. As we round a bend, we spot a raft that, since we've seen so few on this trip, we assume is Matt's and Barb's. After I row a quarter-mile across the river, Barbara reaches for the bow line. Then she whispers, "That's not them!" How wonderful to enjoy so much solitude that we end up confusing any other figures with our trip mates.

To save face, Barbara subtly replaces the line as I hail two men who are pulling on yellow waders. One of these park biologists explains that they're catching and banding rare humpback chubs and razorback suckers, for which these waters provide crucial habitat. Although they're well designed for stability in strong currents, humpbacks disappeared from the Canyon of Lodore when Flaming Gorge Reservoir began to alter flows and lower temperatures. By contrast, the razorback suckers are bottom feeders that require warm flowing water and thus have also declined.

For thousands of years the rivers of the Colorado Basin have flowed cold and cloudy in the spring and warm and clear in the fall. Over these millennia native fish have adapted to the natural rhythms, but dams that change flow patterns and water temperatures have turned specific adaptations into fatal liabilities. Nearly a third of North America's native fishes once swam the rivers of the arid West. Today they're dwindling toward extinction.

It's distressing to recall that the Green River was once a notable fishery. In Powell's time, for instance, "white salmon" (really Colorado River squawfish) growing up to six feet and eighty pounds were common; today squawfish are extinct in the lower Colorado Basin and endangered here in the upper. Since they often die before they reproduce, they now grow to only half their former size. Though dams continue to interfere with their spawning, Flaming Gorge Dam has begun both regulating its releases to better replicate natural flow rhythms and also drawing from the surface of the reservoir, where temperatures run closer to natural conditions.

Soon we enter Rainbow Park, where the colors impressed Major Powell. Here limestones of the Park City Formation top off the blond Weber sandstones, resulting in vivid contrasts. Ravens croak and glide in a lush amphitheater. A blue blur shoots by my head, leaving a squawk. It's a belted kingfisher, and since shallows that offer good fishing are rare, these feisty fishermen are well known for defending their territories.

A mile downriver black-and-buff-winged peregrine falcons plummet from cliffs. One pulls out of its dive just over our heads, sounding as though we've just missed decapitation with a samurai sword. Whereas a hawk's wings extend straight out from its body, a falcon's wings sweep back for speed. Clocked at nearly two hundred miles per hour in a dive, peregrines are the fastest fliers anywhere: even swifts must stay wary. Though at full speed swifts can outmaneuver them, peregrines learn to dive by and then grab one on the upwing.

The peregrine falcon is faring well in these parts. With its inaccessible cliffs, the park provides prime falcon habitat. In fact, falcon eyries here in Dinosaur provide eggs for hatching boxes in—of all places—big cities. From nests on the tops of skyscrapers, peregrines dive toward the street to grab pigeons. Though some of their precious eggs have rolled off windowsills, these urban raptors definitely put on a spectacular show to increase public interest in watchable wildlife.

Below Rainbow Park, the Green River enters a five-mile chute as it races though Split Mountain on a steep gradient of twenty feet per mile, the steepest in Dinosaur. (By contrast, the Mississippi meanders on a gradient of only four inches a mile.) As far back as 1776, early explorers Dominguez and Escalante marveled at how the Green emerged from a "split mountain." Today we understand the effects of faulting that allow a river to slice a mountain in half. With tilted and twisted cliffs, many of them dotted with evergreens, this is a spectacular gorge but its whitewater demands our attention. We buck the waves but skirt the whitecaps to avoid increased seepage in our boots. As its current slows, the Green passes between the tilted sandstone that marks the exit from the gorge, then crosses the "racetrack" of raised rock that rings the south side Split Mountain. Amazing country.

I long for more of the river, but my feet scream for dry socks. In addition to my mentor, the wind and the rapids have taught me a lot; I now feel more ready to undertake big water on my own. This trip has involved risks, and I've made some mistakes, but the river gods have blessed us. We were fortunate to retrieve that oar in Disaster Falls, lucky to boat in the sun and not in the snow, and privileged to have a memorable stretch of river all to ourselves.

Plants of Life

Its foot gently probing a raindrop, a black beetle with orange legs explores the creamy flowers of a tall yucca. A red ant attracted to the nectar follows the beetle inside. Inches away, smaller brown ants speckle the immature seed pods. A lizard lurks under the yucca's leaves, awaiting the ants. So much desert life centers on yuccas and agaves.

In *The Voice of the Desert*, Joseph Wood Krutch describes the complete interdependence between the yucca plant and the pronuba moth. Nothing else can fertilize the yucca, nor can the moth's larvae feed on anything besides its maturing seeds. In fact, this symbiotic relationship is so close that when new species of yuccas evolved, so did new species of pronubas.

As Krutch tells the story, the female moth rests quietly in the half-opened blossoms on her conjugal night:

> While the male, who has already done his duty, flutters uselessly about, she collects from the anthers a ball of the pollen which is surrounded by a sticky gum to prevent its accidental dispersal. After she has collected under her chin a mass somewhat larger than her head, she climbs the pistil of a different flower and into it she inserts her egg tube . . . and injects several eggs. However, she "knows" that if she left it at that, her larva would have nothing to feed on. Accordingly, she mounts the rest of the way up the pistil, deposits the pollen ball on the stigma, and moves her head back and forth to rub the pollen well in.

During her visit the female moth consumes neither nectar nor pollen. Her purpose is simply to generate food for her larvae. In doing this she pollinates the flowers of her food plant. How does she know to lay only a few eggs, so that her larvae will consume only some of the seeds to insure the reproduction of the moth's food plant? Just how did such complex instinctual behavior evolve? I wish I knew.

The agaves, the larger, less-frost-resistant relatives of the yuccas, provide similar habitats for desert life. The agave beetle and the ladder-backed or agave woodpecker, for instance, inhabit this micro environment. After many years spent growing a rosette of huge, succulent leaves, the century plant (*Agave Americana*) shoots up a flower stalk like a giant spear of asparagus. This stalk thrusts up as fast as a foot a day, often reaching fifteen feet, then shoots out candelabra-like branches that burst into waxy golden blooms. These flowers get pollinated by nectar-loving insects, by hummingbirds, and even by long-nosed bats that migrate north when large cacti and agaves flower. After she pollinates the plant, the agave beetle lays eggs in the flower stalk where again the larvae consume most—but not all—of the growing seeds. Agave woodpeckers feed on these larvae, controlling their numbers to insure enough seeds for propagation. Although the number of species is small, desert life involves an intricate web of relationships.

Whereas yuccas bloom several times, agaves die after their initial outburst. Most of the water needed to erect the tall stalk comes from the fleshy leaves. Plants begin to die of dehydration once they begin to bolt. Like leaves on trees, the blades often turn lovely colors as they wither. After the plant dies, the stalk often serves as a nest for the woodpeckers.

Nor have humans remained apart from this complex mesh of life. Yuccas and agaves were essential to prehistoric Indians of the Southwest. In fact, roasting pits are a common feature of archaeological sites dating back almost to the last ice age. The Anasazi made extensive use of these plants for mats, sandals, baskets, thread, and cloth. Some tribes concocted a soup from yucca fruits, while others made soap from agave leaves. The Hohokam of Arizona grew agaves for food and fiber, while the Mescalero Apaches of New Mexico traded in *mescal*, which is Spanish for agave. The Havasupais, Paiutes, and other Indians roast agave stalks in stone-lined pits of coals.

Though sometimes cursed by careless walkers as "Spanish daggers" or "bayonets," yuccas and agaves are hardly hostile. Their leaves are best seen as perfectly adapted to protecting the plant, to preserving moisture, and to reflecting excess light that would otherwise cook these remarkable desert dwellers. To the observant eye, yuccas and agaves serve as hospitable centers for insect, bird, mammal, and human survival.

9.

Kolob Backcountry: Paradise Found

Canyons, through which great rivers roll onward to the ocean, and whose walls rise up so high as to shut out the glare of day . . . all pale before . . . the Mu-kun-tu-weap valley of the Virgin River in southern Utah . . . whose brows confront the sky . . . dazzling in their barbaric splendor of color, their scenes of magnificent disorder.

—H. L. A. Culmer, *The Scenic Glories of Utah*

Among the spectacles of the Colorado Plateau, few surpass the high plateaus of southern Utah and northern Arizona. Here in addition to the usual beauties offered by the red-rock country, one finds a crowning glory: large trees. The luxuriant foliage that drapes the rocks here at Kolob Terrace rivals that of Zion Canyon, itself a plant lover's nirvana. Like Mayan pyramids overgrown with greenery, the burnt-orange sandstone bluffs round into the domes so characteristic of Zion. In most of the park, cliff tops appear blond where water has dissolved much of the lime and iron that cement the quartz crystals, but here the Navajo sandstone bluffs retain more of their red pigments. Our route toward Kolob Arch skirts this balcony of red and green. From all we've heard, this should be one of the great hikes in the Southwest.

Major geologic factors explain these spectacular walls and domes. In essence, winds piled up vast dunes that were flooded, covered, lifted up, and eroded down to become the Colorado Plateau. When the sands accumulated, this area lay even with present-day Central America. Since that time, tectonic plate movement caused the continent to drift two thousand miles northward. Despite its origin in the Age of Dinosaurs, the Navajo sandstone contains few fossils because few plants and animals lived in the hot desert sands.

The "Great Sand Pile" reached its highest point here in southwest Utah, and the resulting Navajo sandstone stands fully two thousand feet thick in Zion Canyon. But upheavals that began about thirteen million years ago lifted this rock several thousand feet higher yet in the Kolob section. The upthrust of this colossal block exposed it to the elements that carve the domes, towers, slots, and ruddy walls along this trail.

We're skirting the Hurricane Fault, the west-facing escarpment that runs for almost three hundred miles from the Wasatch Mountains to Grand Canyon. Viewed from the air, this long cliff marks the west side of the Grand Staircase. Perched behind these sandstone bluffs, the swaled, corrugated Kolob Terrace represents a middle step on the stairs. Above this shelf, the Gray Cliffs of Cedar Canyon and Pink Cliffs of

Bryce Canyon reach the top landings. Downstairs from Kolob Terrace lie the White Cliffs of Zion, the Vermilion Cliffs of Kanab, the Belted Cliffs of Colorado City, and finally the Kaibab Limestone bluffs of the Grand Canyon. As it scales the six thousand feet from the Colorado River to Bryce Canyon, the Grand Staircase reveals nearly two billion years of the earth's history.

Runoff from Kolob Terrace not only carves the canyons we'll explore but nourishes a unique blend of plants from the Rocky Mountains, the Great Basin, the Sierra Nevadas, and the Arizona desert. Considering the fact that much of the park is bare rock, this variety is amazing. Zion is a place of specialized adaptations to micro niches; standard zones often don't apply because micro climates vary widely in a short space. A cactus may thrive on a sunny ledge a few feet from the lush hanging gardens of ferns and monkey flowers, the special habitat required by the Zion snail, a species found nowhere else.

We parked at Lee's Pass, named for John D. Lee, a complex and controversial figure who secretly used this route on his way to Hop Valley and beyond. Since the Mormon church commissioned him to explore it, Lee knew southwest Utah as well as any white man. His well-known troubles began in 1857 when Paiute Indians harassed over a hundred non-Mormon emigrants camped at Mountain Meadows, thirty miles west of here. When the immigrant party circled its wagons and repulsed the Paiutes, inflicting casualties, Mormon leaders became fearful. Would the attack activate the U.S. Army, which was already poised to invade Utah? Would the Indians turn on the Mormon settlers?

Following instructions, Lee and others first offered to help the emigrants, then led them into a trap: the Mountain

Meadows Massacre. The Mormons fell on the men as the Indians attacked the women and children. Weeks later wolves still chewed on the bodies. The distinguished Western writer Wallace Stegner rightly described the tragedy as an outburst of "hatred and misunderstanding that had been building for a long time."

Excommunicated and forced into exile, Lee often took remote routes such as the trail we're hiking. After he'd sequestered himself for years at Lee's Ferry on the Colorado River, he was eventually tried and brought before a firing squad. To his dying breath Lee contended that Brigham Young "sacrificed me through his lust for power." Though Lee, a multiple murderer, never denied his involvement in the massacre, he also bore the guilt of a tribe that had disowned its shadow.

A thousand feet above us, the last of the rounded Five Fingers marks the western reach of the red rock plateau. Like most drainages that incise Kolob Terrace along cracks in the rock, those between the fingers run east and west. The trail continues south, first along cottonwood-lined Timber Creek, then heads through pinyon and juniper country.

Butterflies probe the brilliant orange blossoms clustered on a handsome bush, its dark-green leaves set against a background of gray sage and red rock. Anchored to deep, drought-resistant tubers, butterfly weed (or orange milkweed) blooms in warm colors that make it attractive for landscaping. These butterflies face a risk, however, for pollination in the milkweed family can prove fatal. Rather than floating freely for grain-by-grain transfer, this pollen sticks together in a waxlike mass. With great strength relative to their weight, visiting insects must free the waxy platelets of pollen or become trapped in the flower. When things work well, their pollen-laden feet inadvertently grasp the protruding pistol of the flower. Mass fertilization ensues.

With its diversified flora offering nectar throughout warmer months, Zion attracts several varieties of butterflies. Buckeyes, small but strikingly colored butterflies with blue eyespots, often dash after others of their kind and then ascend in a whirlwind. California sisters resemble white-striped admirals except for the pumpkin-orange dots that glow like tail lights. Azures, painted ladies, tiger swallowtails, and common sulphurs (whose color probably suggested the term "butter fly") all add to the already diverse palette.

As La Verkin Creek enters from the east, we look around the historic *Corrales*, where for decades ranchers paused with their herds. Feet and hooves have compacted the soil, leading the Zion Natural History Association to advise against camping on such trampled areas. From here the trail heads into the complex of verdant buff-and-salmon canyons that we'll explore.

Like a painted bird on a Christmas tree, a Western tanager swoops from a treetop, flashing its red head, yellow-orange body, and black tail. Tanagers are birds of the tropics, reaching their zenith on the lower slopes of the Andes where several species may feed from the same tree at the same time. In the summer males flaunt their colors for breeding, but in the fall, for protection during migration, they molt into the dull greenish-yellow of their mates. With their bursts of flamboyance, tanagers bring the tropics to temperate North America. Bird lovers now recognize the crucial importance of preserving winter habitat for "our" summer songbirds.

Skies darken as clouds engulf Timber Top Mountain. When the trail climbs into conifers, drops begin to tingle my scalp. While we munch carrots under a spreading ponderosa, the drizzle wanes into mist. A cloud burns off as Gregory Peak, named for pioneer geologist Herbert E. Gregory, blazes in the late-afternoon light.

Campsites nestle in shady groves. No sooner do we put down our burdens than a rock squirrel tears into my pack. We yell, make coarse gestures, and even charge the marauder, but it doesn't scurry far. To prevent further problems, Barbara guards the packs while I drift down to the stream. Its banks lined with bright green horsetails, it rushes brick red after the rain. A Say's Phoebe, gray with a sharp beak, shoots out to snap a mosquito. On the banks grow plants commonly found in the East, such as Solomon's seal and black raspberries. Just downstream a buck kinks his neck to scratch his steaming back with the tip of his antler.

This area is well known for wildlife. When trapper-traders William Wolfkill and George C. Yount traipsed through southwest Utah in the winter of 1831–32, Yount marveled at what they saw in the Virgin River valley: "The elk, deer, and antelope, driven from the mountains by the snow and piercing cold, were basking, with their frolicsome fawns, unaware and unintimidated by the sight of man. They would flock around like domestic sheep or goats, and would almost feed from the hand."

Where there was plentiful prey, there were also predators. Unfortunately, however, hunters, trappers, and ranchers had exterminated the Mexican gray wolves well before establishment of the park. Mountain lions, on the other hand, proved more difficult to kill. Their preference for fresh meat made them difficult to poison, and they hid on ledges where they could evade a pack of howling hounds. Much as a house cat can reach a counter, cougars can spring up cliffs several times their height.

These magnificent cats now prowl Kolob Terrace, but, because their territories are so large, they often stray from the park. When cougars prey on livestock and a rancher reports the loss, the Utah Division of Wildlife traps the cougar for release in areas needing predators. But other ranchers prefer to "bag their lion" like a Teddy Roosevelt—or find it profitable to guide outstate trophy hunters whose challenge amounts to hitting a large target crouched in a small tree. Cougars are defiantly wild, and there are people who need to destroy any creature that defies human domination. Perhaps Captain Ahab, who hated the whale he couldn't kill, didn't go down with the *Pequod*.

Mountain lions have also faced accusations of killing off the bighorn sheep here in Zion. Actually, though, there were additional causes. When the deer population expanded, bighorns were less able to compete, so their numbers dwindled. In addition, reintroduced bighorns have multiplied slowly in Zion because of diseases contracted from domestic sheep. Small parks such as Zion and Bryce Canyon present special challenges. Both have lost about a third of their original species, and the reintroduction of mobile species such as wolves alarms the locals. It's extremely difficult to restore full diversity in a complex ecosystem that's been evolving for thousands of years.

In the morning we face the rock squirrels again, so we stuff our food into a sack and suspend it from the end of a bough. But the sack's weight bows it down and squirrels jump *that* high to reach bird feeders. I know—I've been outsmarted by them before. Finally we select a stronger branch, clean any crumbs from around our site, and set off to play.

Our first stop is Kolob Arch, at 310 feet possibly the largest in the world. Despite its size, the arch itself is not overwhelming because it sits so high up the cliff. Unlike many arches on the Colorado Plateau, Kolob is not essentially the work of percolating water. Instead, it is a free-standing arch, one formed through exfoliation. Where the weight of the rock creates bulges along stress planes, cracking and peeling loosen surface slabs that slough off during freezes. As time goes on, expansion and contraction accelerate this process. Much like Great Arch near the Zion Tunnel, Kolob has exfoliated itself deeply into the cliff, finally causing its ceiling to fall in. Because Kolob is open at the top, it does not protect the cliff beneath it from eroding still more rapidly than before.

Few hikers continue past Kolob Arch. Although the lovely trail to the Arch is graced with pink phlox, scarlet penstemons, lavender shooting stars, and other wildflowers, it scarcely hints at the marvels just beyond. We follow the stream that has, for only a few million years, abraded this soft sandstone like a continuously grinding belt sander. Yesterday's shower speeded up the belt and added new abrasives.

For another mile past the arch, these exquisite narrows remain pristine, free from bootprints. Far above, cross-bedded planes sweep through long arcs on the canyon walls. Fresh fractures appear clean and buff against the older, darker surfaces. Seeps spot these cliffs to reveal where moisture percolates through the porous sandstone, making them resemble the Redwall limestone bluffs in the Grand Canyon. However, these sheer walls rise over a thousand feet in two stages, separated by a distinct seep line where water reaches a harder layer which it follows until it oozes out. Strong enough to support itself on a cliff face but soft enough for streams to chisel, Navajo sandstone forms absolutely magnificent cliffs.

Well up the sheer face, a narrow shelf supports a band of trees. Tall conifers too high up for even a tree hugger to identify look like props in a museum showcase. The upper wall frames lovely water streaks, long stripes that accentuate the curvature. This second face, too, is crowned with twisted trees standing defiantly against a blue slice of sky.

These glorious walls glow like a pumpkin in the mid-morning sunlight. In such narrow canyons the reflected overtones can render the colors richer. Flavored by this tangerine light, mist drifts from beneath an overhang where lacy maidenhair ferns decorate the walls. Like Weeping Rock in Zion Canyon, this shallow cave drips tears that wobble in the air like soap bubbles.

The air seems slightly tinted by iron, full of grains from the sandstone cliffs. This is broad daylight turned palpable, not photons simply striking and bouncing off a surface, not rock merely absorbing some wavelengths of light to impart color, but sublime luminescence imbued with the spirit of matter. This beauty animates me, dissolving any feelings of estrangement into the sandstoned air. I react to illusory appearances, momentarily disregarding the hard-rock realities that lie beneath the gorgeous surfaces.

Far overhead, their wings backlit by the sun, hundreds of swallows twitter and cavort. Cliff swallows winter in Paraguay and Brazil, flying thousands of miles each way and nesting in huge colonies. These colonies serve as communication centers where unsuccessful hunters watch successful ones feeding their nestlings, then follow them to the best places to grab bugs. However, parasitism also characterizes swallow colonies. Laying eggs in neighboring nests is common, even rampant. Nor is this the result of confusion among similar nests, for some females wait until the owner is distracted, then pop out an egg in as little as fifteen seconds. Exuberant fliers that they are, swallows understandably prefer soaring to brooding eggs.

In the creek a slate-gray water ouzel, or dipper, bobs on rocks to remind me of how I first beheld ouzels under a bridge in Yosemite. In his *The Mountains of California*, John Muir sang paeans to his favorite bird:

> He is a singularly joyous and lovable little fellow, about the size of a robin, clad in a plain waterproof suit of bluish gray. . . . In both winter and summer he sings, sweetly, cheerily, independent alike of sunshine and of love, requiring no other inspiration than the stream on which he dwells.

Water ouzels dive into pools where they hunt insect larvae. And they aren't fooled by caddis fly grubs that cover themselves with sticks and stones, either.

Where huge slabs of sandstone clog the slot, we pull ourselves up and over, each time marking the best route to climb back down. After running nearly straight, apparently along a north/south joint, this slot opens into a spectacular amphitheater. Black-and-white streaks decorate an enormous stone knob. Tufts of grass and moss—plus a limber pine seedling, its boughs like bottle brushes—all cling to the peeling rock.

How appropriate that, as far as we can tell, this magical alcove doesn't have a name. Sequestered behind those slab barriers, this is a magical place that few people see. Perhaps it's better off unnamed, for blank places on the map are less likely to risk the overuse experienced by some of Zion's backcountry, such as The Subway.

On the way back I freeze. Just a yard ahead, its head buried under loose rock, lies a colorful serpent. Its strikingly colored body, larger than my thumb, looks almost enameled. Barbara cautions me to stand back, which I do, though I'd prefer to see as much as possible. Yes, I definitely do have a way of letting beauty blind me to danger. Then I recall that "Red and yellow, kill a fellow. Red and black, venom lack." Rings encircle the body, but the red ones do not meet the yellow. Red rings are bordered on either side by black. This is a harmless Arizona mountain kingsnake, a constrictor of lizards and other snakes. I linger for a closer look.

My skin feels sticky as I recline on a flagstone, legs in a pool, snacking on limp carrots and stale peanuts. Mmmm, good. Gauzy-winged damselflies dance with Tinkerbell wings in the dappled light. Their bowed bodies glowing a phosphorescent blue, they often perch near sunny spots so that small flying insects wandering into the sunshine are easier to see. Several damselflies light softly on stream grasses beside my canteen. After a mating display of arched abdomens, a male fastens his tail behind a female's head. Later she will curl her abdomen around to meet with his, suggesting a most holistic union of mind and body.

A small black water snake two feet long wriggles into the pool. Head held high, it glides up and rests on a rock ledge. It remains calm, perhaps more so than we do, ignoring both us and the damselflies. It then swims downstream, much at home in the gentle current.

Buzz. Leaf litter flies as two antagonists fight to the death. A yellow jacket is assaulting a large digger bee, its wings a blur in the dust. Gradually the yellow jacket pins the bee, going at it with chewing mouth and stinging tail. While its jaws bite the bee's head off, its stinger repeatedly stabs the bee's abdomen. The decapitated bee's wings buzz intermittently. What's the conflict here? This yellow jacket may want the bee's burrow, since both nest in the soil. It's more likely, though, that this yellow jacket is a sterile female worker seeking protein to feed larvae. After an episode like this, it's difficult to understand why most bees live alone. There's greater safety in numbers, but bees that live in hives must pay a price for their greater security.

As I stare at this savage scene, I'm both shocked and fascinated. When nature seems as beautiful or as suited to human needs as it does right now, I forget that it can also be so "red in tooth and claw."

Back at camp we're in for another shock; the rodents have struck again—and I've been outsmarted once more. Familiar debris litters the ground near our stuff sack. The varmints have chewed through the cord to drop our sack of food. Though it's humbling to be outwitted by rock squirrels, the loss is more serious. Without our dried dinners, peanuts, granola, and dried apples, we're short on food. Barbara wants to hike out; I argue that the little beasties didn't take our coffee, oranges, peanuts, or brandy. We should be able to get by on these and then hike out early. When I promise her a rib eye at Milt's Stage Stop near Cedar City, she agrees to stay.

After the cleanup we set out for Hop Valley. A half mile down La Verkin Creek we come to a fork. Ringed by golden columbine, a sizable spring gushes from the hillside. The streamside trail follows La Verkin and Willis creeks past more springs into slot canyons, some of them classic narrows. Surrounded by Wilderness Study Areas, this is one of the wildest areas of the park, one where large pines and firs, rushing waters, waterfalls, pools for bathing, and solitude are the rule.

The Hop Valley trail is also spectacular. From the creek the trail climbs into a thick grove of maples and box elders. Domes and towering cliffs rise in nearly all directions. The cross-bedding characteristic of wind-deposited sandstones often guides erosional forces, determining by their tilt whether the rock will become a dome or a cliff. Where the Navajo sandstone is exposed on these great rock faces, the curvatures of ancient dunes define the sweeping contours. The sandstone domes of Zion present the crests of petrified dunes, frozen in time. Except for Gregory Butte, which balds at 7,700 feet, these rounded cliffs are trimmed with green.

Hop Valley first presents wide, rolling terrain covered by scrub oaks and ponderosas, then it changes to broad meadows with blue lupines, or bluebonnets. Since lupines are poisonous to cattle, their numbers have probably increased as decades of grazing have reduced more palatable plants. Other changes have resulted from protracted grazing. Sage, pinyon, and juniper often invade lands denuded of their natural grasses. Farther on, where slump-and-slide dams once backed up the stream to create a lake, the valley widens. Its floor resembles a golf course, complete with sand traps and fairways. An eagle sails into the valley, stirring the ravens.

Most of this long valley is an inholding owned by Bud Lee, a descendent of John D. For years Bud has run cattle in these verdant meadows from May through October, but the cattle also stray into the park, where they do damage. Private inholdings disrupt national parks when their owners insist on uses that conflict with park values. Why, one wonders, can the government condemn private land for highways, railroads, power lines, and pipelines but not for unified national parks? Here in Utah, at least, one answer is that the Park Service has felt intense hostility from locals, especially ranchers, and has tried to appease them.

On the final trail back, other campers tell tales of woe. They thought they'd be clever and hide food in their tent, but they too were outsmarted. Squirrels chewed into their tent, found their food and fought over it, smearing blood and excrement. Normally rock squirrels eat acorns plus the seeds of currant, cactus, and yucca, all of them plentiful in this area. Did these rodents become dependent on human handouts? Or did they overpopulate, leading them to become excessively aggressive? When animals lose their fear of humans, could they turn their residual anger on us? Naturalists often remark that in areas frequented by humans the wildlife comes forward, whereas in truly wild areas it runs the other way. These squirrels parallel the fearless deer that one finds around many campgrounds. Contrary to "the Bambi syndrome," such seemingly tame, cute animals are not as benign as their liquid eyes might suggest. On more than one occasion they've chewed off my bootlaces or carried off my hat, both of which cause problems when you're camped in the backcountry.

The other campers decide to hike out, but I forget to make an offer for what remains of their food. Damn! To ignore the hunger pangs, I imagine spending time to select just the right twigs for a small, low-impact fire and whittling as I toss shavings on the fire. But no, this is a national park where no backcountry fires are allowed. Besides, freed from cooking we have more time to observe everything. If full vitality involves being totally present, nature delivers like little else.

But writing offers full involvement too—should I watch the woodpeckers or scribble in my journal? New England poet and essayist Donald Hall mused on contentment and "absorbedness":

> The hour of bliss is the lost hour. . . . I lose the hour—
> inhabiting contentment—in my lucky double absorption
> with work and with land. At the desk, writing and trying to

write, I do not even know that I will die. The whole of me enters the hand that holds the pen that digs at word-weeds, trying to set the garden straight.

For total absorbedness one watches, writes, reads, gardens, converses, makes love, or whatever.

Toward dusk, as clouds move in, we exchange massages to feel less like we've been sent to our rooms without supper. Rain infuses the woods with an immense throb of energy, pitter-pattering the tent with insistent rhythms. We snuggle into our bags, drifting slowly toward sleep, relaxed but excited. I sense along with Thomas Merton that rain "reminds me again and again that the whole world runs by rhythms I have not yet learned to recognize."

Once the last drips from the oak have ceased, droplets of light ooze through the nylon from a sky of pearls. Outside, the Summer Triangle is swimming beside the ghostly, gauzy Milky Way. As the skies clear fully, I hear the stars hum and listen to the music of the spheres.

Living in a city, where light pollution limits our view of the cosmos, I'd forgotten the splendor of the night sky. Here, thankfully, we're a hundred and fifty miles from the gaudy glare of Las Vegas. Air pollution also clouds our capacity for wonder, our ability to commune with a universe that is so much more than "virtual" dots on a screen. Today most people settle for a surrogate sky in a planetarium where viewers gasp in astonishment as the dull city heavens burst into their former brilliance. Under an empty sky or one filled only with blinking lights from airplanes, our engagement with the stars that so fascinated our ancestors lapses into disconnection with realms beyond our own.

I awaken to thunks on our tent. These are just too

regular, so I suspect that a squirrel is dropping pine cones. Perhaps I've got rodents on my mind. We slug down coffee and tear into the one last orange, thinking we'll need a double caffeine and fructose buzz to cover the seven miles. With no delay for washing dishes, we're off by eight. Over a thousand feet above us Gregory Butte gleams in the early light, tiger-striped by last night's rain. An evening primrose, a common night bloomer, looks shriveled after its night of glory.

Raindrops bead the knee-high golden buckwheat beside the trail. This is the local version of the wild buckwheat appearing as sulphur flowers and umbrella plants in the Rockies. Throughout the West, wherever differences in moisture, salinity, soil type, exposure, and animal foraging occur, different species of buckwheat adapt to specific eco-niches. Perhaps the most dramatic adaptation to aridity is *Eriogonum inflatum*, the bottle plant that stores water in its bloated stems.

Beside the buckwheat, sphinx moths mate in the moist shade. Their colors are muted but bold, their forewings dark brown with a beige band, their hind wings mostly pink. Named for their large caterpillars that rear up like a sphinx, these remarkable insects are sometimes called "hawkmoths" because of their swooping flight or, more appropriately, "hummingbird moths" because of their hovering movements and rapid wingbeats. Unlike most moths, their antennae aren't feathery to locate food in the dark. Instead, their eyes are highly developed to spot bright-colored flowers during the day.

Sphinx moths lack hearing organs, possibly because they wouldn't be able to hear above their whir in flight. Stout-bodied and strong-winged, these insects buzz from flower to flower, often at dusk. As they hang above a flower, their pink color visible through the blur of their wings, they grasp the petal with two legs while their long proboscis uncoils to penetrate the flower as if in a passionate embrace. To fuel their high-energy metabolism, they require nectar with a lot of sugar. Burning all this fuel makes it necessary for them to release heat, which they do by perspiring from their wings and by circulating extra air through their respiratory system. Who could dismiss these marvels as mere bugs?

The sphinx moths don't amuse Barbara, though: "You're an interesting guy to hike with, Paul, but not on an empty stomach." Like a horse galloping for the barn, she heads for her oats, non-stop. Pulling on blinders, I force myself to bypass the kingbird on the branch, the butterflies on the pleurisy-root bush, and the baby grouse, almost invisible among the rice grass and sage. A curious coyote probably gives us more time than we give it. Barbara remarks that she'd complain about her sore feet if they didn't keep her mind off her belly. Hunger hits me too. I feel weak, and not just from the heat. After groaning up the last grade, we heave our packs somewhere near our van.

While I drive, she scavenges like a rodent for crackers, bits, pieces, crumbs. At Milt's Stage Stop we wolf down rib-eye steaks like ravenous carnivores. But my conscience nags. How can we object to overgrazing and then eat beef? Sure, we human critters are omnivores so it's natural for us to eat meat. Everything considered, though, I vow to consume even less red meat. Starting tomorrow. Over coffee, we rave about these glorious higher canyons of Zion. Their altitude and moisture keep them cooler, and the extra water supports more varied and prolific flora and fauna than in lower desert areas. When the Mormon pioneers named this Kolob area after the star nearest Heaven, they weren't far wrong. This might make a good place to live. . . .

Three years later we moved to Cedar City, only eighteen miles away.

Interlude

A Sentimental Homecoming

Located just off the interstate, the Taylor Creek Trail in Zion National Park attracts many casual visitors, but Ray Fife certainly wasn't one of these. Under his white cowboy hat he gazed around as though in search of something he'd lost.

"I can't believe how everything's changed," he mused. "The canyon's grown up so much—it was much more bare back then, after the lumbering, with all the grazing going on."

Now a retired physician, Ray recalled his often solitary adolescence in this canyon. His father Arthur, a professor at the Branch Agricultural College in Cedar City, loved these red rocks so much that he endured twenty miles of bad roads to live here, where he ran cows, pigs, sheep, and goats.

"Ranching was hard work. The flash floods were real frustrating 'cause they washed away the fencing you'd just put up. So I didn't appreciate the beauty when I lived here as a teenager. Guess I wanted to be with my friends back in town. But we did have excitement out here at times, like when my dad brought some students into one of these box canyons and a mountain lion sprang just over their heads on a ledge, twenty feet away."

Those who were lucky enough to live amid natural beauty while they were young may not have appreciated it, but such early immersion usually forges a bond that enhances their relationship with nature later on. Ray and I sauntered so he could talk and stay within his breath. He gestured toward Ash Creek where John D. Lee, plus other early settlers, came to cut hardwood for tools. As a result, there still isn't much ash growing along these creek bottoms. Ray reminisced about the ruins of a cabin reputedly built by Lee, then pointed out some twisted rock beds known as the Kanarra Fold.

"And don't let me forget to show you the conglomerate—my dad really loved it." Ray told how his cousin Louis Fife, an Iron County sheriff, tracked down the last of the Old Western outlaws, Jack Weston, only to get captured himself. With the help of his partner, Weston handcuffed the sheriff to a juniper and left him for dead. Despite wounds and fatigue, however, Sheriff Fife managed to work his way up and over the nineteen-foot tree. Undaunted, he later single-handedly recaptured the outlaws.

As the gentle doctor pushed back his Stetson, he marveled at Tocupit Point, the lofty prow which divides the forks of Taylor Creek. Which of the Fingers of Kolob is it? I wondered.

"Behind the point is the Temple Cap, the brick-colored layer above the sandstone cliffs. Above it, the Carmel limestone is full of fossils," Ray recalled. After all these years he still knew his rocks. As we arrived at the cabin he helped his father build in 1930, his voice quavered.

"We hated to lose all this when they made it a National Monument, but I guess it's better off that way." As he choked up, I reached for his wrinkled hand before Ray ambled toward the cabin he'd helped build, sixty-two years before.

Throat loosened, I slogged down the creek bed toward the huge rockfall, dam, and pond. I so wished Ray could see this. At times such falls have blocked streams to form lakes in the canyons. Known for how it allows precipitation to percolate, Navajo sandstone can absorb moisture and add weight. Or when the lime that serves as glue dissolves, huge blocks can disintegrate in a fall. In just seconds, rock crumbles into sand. But it was neither heavy rains nor wedges of ice nor an earthquake that caused this cliff to rupture, dropping large chunks nearly a thousand feet; apparently it was expansion in the summer heat. For several days locals across the valley in New Harmony watched a cloud of orange dust that hung over the Five Fingers of Kolob.

The next year a flash flood carried vast amounts of this freshly freed sand, almost closing the interstate. The sand and water scoured the Middle Fork, filling in the pools that formerly hosted so much acquatic life. Fewer canyon frogs will be bleating along here for the next few years, but already yuccas and other plants that did not grow on the canyon floor are taking hold on the new dune. Nature may reach equilibrium, but it also reacts to disturbances, including lumbering, grazing, rock falls, and flash floods.

On my way back I wondered if Ray had ever read Everett Ruess, "the vagabond for beauty" who gave up everything, including his life, to inhabit the red rocks he loved. Ray would probably resonate with Everett, who was young when he was. Everett's heartfelt letters might touch him deeply. But when I arrived back at the trailhead, Ray was gone.

There was so much we could have shared, much like father and son.

10.

Big Water on the Green

*Floating the rivers takes you through the land, not merely
over its surface. Entering a canyon is akin to entering the
living body of the earth, floating with its lifeblood through
the arteries of veins and rock, turning your perceptions to
the slow pulse of the land, single beats of river current
marking the steady rhythmic changes geologic.*

—Stephen Trimble, *The Bright Edge*

"Just the two of you goin' down?" asks the kayaker. Funny he
should inquire, for this is our first run without another boat.
Once into the flow, we'll do the full eighty-four miles. If we
flip our boat, break a leg, come down sick, or get stuck on a
mudbank, we can't expect much help, though we won't
decline any offers.

With the spring runoff cresting, the Green River is rolling along at a hefty eleven thousand cubic feet a second. Running a river of this size promises freedom but also entails giving up control. Big river, small boat. It's not surprising, then, that Barbara and I appreciate the camaraderie that enlivens the launch at Sand Wash. When our heavily loaded boat sticks in the muck, other rafters grunt and heave until it breaks free. I stroke into the flow, kicking water jugs from underfoot. Gusts from downstream sparkle on the otherwise glassy surface as the river heads for one of the West's great wildernesses—Desolation and Gray canyons.

Before long the rocks become impressive. On the left loom "The Wrinkles," jagged bluffs that rim Nutters Hole, a colossal amphitheater extending fully two miles across. The "Gothic Cathedral," named for the buttresses that section these cliffs, appears around the bend. These Book Cliffs will rise as the river penetrates a mountain range where, over geologic time, the Uinta Basin has risen while the river has eroded down its entrenched channel.

On the right, a clump of cottonwoods and tamarisk marks the mouth of Nine Mile Canyon, once a center for the Fremont Indian Culture. It's ironic that a native people would bear the name of the Euro-American explorer who arrived more than five centuries after they left. That aside, the rock art is outstanding up this canyon, a place the BLM calls "the longest art gallery in the world." Since the majority of the art depicts game animals, we might infer that this oasis served as a prime hunting ground for the Fremont people. The remarkable verdancy of Nine Mile Canyon also supported beaver, for as early as 1822 it had become a primary trapping area for the first white men to paddle the Green River.

This greenery also led early cattleman Preston Nutter,

who ranged massive herds from Colorado to Arizona, to establish his main ranch here. Later, in a move than would foreshadow Western history, Nutter began to control vast areas by holding the grazing permits to public lands. Besides all the cows, Butch Cassidy, "Gun Play" Maxwell, and other outlaws roamed Nutter's spread. Since outlaws were usually cowboys by trade, between robberies they often resorted to rustling. Rather than see his cattle stolen, Nutter shrewdly hired outlaws as cowhands.

The grayish-tan cliffs of Summers Amphitheater, named for Jack Summer of Powell's first expedition, resemble the semi-circular colonnade of St. Peter's square but on a grander scale. Here in Desolation Canyon the rocks become Wasatch limestones and shales, the same strata that, tinted with pastel pigments, outcrop so splendidly at Bryce Canyon and Cedar Breaks in southern Utah. Because the strata tilt north and the river cuts down as it runs south, we're moving back in geologic time at the rate of about a million years a mile.

An intermittent drone announces giant motorized rafts, each towing smaller boats loaded with five people apiece.

"Man, that outfitter is really packin' 'em in," Barbara remarks.

"Yeah, must be fun to listen to a motor and inhale fumes for a couple days." When this flotilla stops, it dumps thirty visitors onto a small beach where many stand in one another's footprints.

Around the bend is an alluring beach under a clump of box elders. This site offers logs for lounging, a weathered plywood table, and, above all, plenty of full shade. Heads cocked to look us over, brown lizards hang from the undersides of the branches. Barbara ties us to a tamarisk bush sporting its full lavender plumage. Marsh grasses waver, billowed by the afternoon gusts. Red-winged blackbirds shoot back and forth along the river's edge before landing atop swaying willows. The females, streaked in browns, are camouflaged for sitting on a nest in the reeds, while the males flaunt their red epaulets in territorial displays.

Enchanted by the flow of the river and the play of the light, we sit in rapt silence. When I sigh that we'd better unpack, Barbara looks shocked.

"We can't camp here—it's illegal anywhere near Rock Creek," she protests.

"But this isn't a creek; it's a dry wash."

"Don't quibble over semantics."

"Do we really want to search for another campsite? It's hot, I'm tired, and we're both already looking like broiled lobsters."

"Look, Paul, rules are made for reasons."

"Yeah, usually. The reason is to protect high-use areas, right? I guarantee that we won't degrade this campsite."

"Let's flip a coin, then."

"OK—got one?"

"Don't be silly."

Paging though the *Desolation River Guide*, Barbara discovers that Rock Creek enters almost thirty miles downstream. This is Rock House Bottom.

We explore the huge slab that created this rock "house," or shelter. Heat still glares from the broken rock. The petroglyphs nearby are even more interesting: squiggly lines leading to the eye of a needle; elk, bear, and bighorn rams, one with an arrow in front of his nose; and two dynamic figures facing each other as they duel or dance.

Toward twilight the Green once again delivers its charms. We linger in river chairs watching the orange-and-lavender light tint the grayish cliffs. Like bony fingers raised in protest, aged junipers reach skyward atop the windswept cliffs. Underfoot, the sand is warm and soft. After being encased for too long, my bare feet revel in their nakedness. At dusk the Rocky Mountain toads begin to hop about, their chins blasting through the parched grass. I inch my feet forward, wanting to watch but not to squash any toads.

As darkness falls, a nighthawk cries as it hunts for insects. Large bats swoop right under the low-hanging boughs just above our heads. Smaller pipistrelle bats with fat bodies and stubby wings flutter just above the grass. The cool moist air draws pipistrelles, for they can't afford to lose much moisture to perspiration. With fur as blond as a marmot and as soft as lamb's wool, pipistrelles hardly fit the stereotype of vampirism. In fact, they don't grab bugs with their teeth at all but instead catch them in a pouch located behind their wings. Maligned by stories of sucking blood or tangling in human hair, bats have been victims of speciesism—the habit of ranking one creature over another.

The next morning a loud, scolding squawk awakens us. When I stumble out for a look, the bird disappears. Probably a chat. These yellow-bellied tricksters can be secretive, I've found. But when I peer into a bush, I do get to see this evasive bird. She's a chat, all right, and she's sitting on her nest. However, her alarm call resembles a cluck, not a wolf whistle. For many river miles I've mistakenly thought that these birds were rebuking me with their sharp calls. You can't take strange birds too personally.

Exposed by a drop in the water level, our raft sits stranded on a mud flat. As the river receded, a family of field mice has nested under the boat. When I lift it, they scatter, running and swimming madly along the waterline. How will these mice ever find their way back together?

Four American avocets, their skinny bills curved upward for underwater probing, dredge the shallows for small crustaceans. The two younger birds flaunt cinnamon feathers on their necks, while the two adults are crisply black and white. These avocets are feeding in ideal backwater habitat—shallows with unobstructed views for optimal protection against predators. Should a sparrow hawk make a move, the avocets are ready: while one screams the alarm, others gang up on the intruder to drive it away. Running north and south as it does, the Green provides an essential flyway for many migratory birds.

With Barbara at the oars, I ease myself over the tube, reluctant to face the shock to my torso. Riffles lap against my chin, and my hands become invisible just below the surface. As it bobs on its way, a well-chewed, barkless piece of beaver wood becomes my companion. When Barbara leans into her strokes, we come together. Arms rubbery, I flounder aboard, reconnected to the river through ritual immersion.

Before long the river begins to run faster. Between the Sand Wash and here at Jack Creek, the Green has been dropping only a foot a mile. From here to its emergence from Gray Canyon, however, it will drop six feet a mile. According to the river guidebook, the rapids are predictably situated where side canyons dump broken rock that narrows the channel and further increases the speed of the current. The actual hazards are less predictable, though, since they change with each flash flood that dumps debris. In a river this large the main hazards are boulders or cottonwood stumps; the current rolls the smaller stuff out of the way. Underwater, often disguised by chop and froth, giant boulders raise slightly elevated pillows. Behind them, often unseen, lurk the sunken holes so dreaded by boaters like us.

As the river roughens, Barbara wants me back in the rower's seat. In this first whitewater we sharpen our eyes and reestablish our commands from spotter to rower. Though, like many couples, we share our own slang, we don't talk river lingo every day. Then I face a real problem. I've made the mistake of rubbing sunblock on my brow where sweat has caused it to run into my eye: the sting is so intense that I can barely function at the helm. Steer Ridge Rapid, our first big challenge, lies somewhere ahead, though we don't know just how far. The maps show five islands two miles above Steer, but what constitutes an established island and what's a new sandbar?

My world soon becomes a watery blur. We're getting swept, semi-blind, toward Steer Rapid—we've got to find a haven. But rapids are coming with increasing frequency and we can't pull off in a rapid. Barbara spots a clump of trees. I pull hard but the current sweeps us by the landing. My eye is stinging worse, but Barbara can't take over here. She scours the banks as far downstream as she can see, but spots no landings. I don't dare drop the oars long enough to rinse my eye.

Finally Barbara locates a cobbled shoal; the oars skid on the stones. She jumps out, rope in hand, and plows through the current while I writhe and grope for a canteen. As I stagger ashore, Barbara sits me down, tilts my head back, and floods my eye. Within a long minute the stinging finally begins to subside.

We're looking for a site with both shade and a flat place to sleep. One cottonwood looks promising, but bleached twigs bristle on the ground beneath it. Another area looks grassy but proves to be a willow thicket. Sand verbena and sweet alyssum whiten the buff-colored sand. I scramble up a bank, but here the trees are too small to offer shade, and cactus bristle from the ground. A hundred meters downstream Barbara's found a diseased, broken-down old cottonwood that yields some shade. This is nothing like our first site, but it'll get us through the night.

In dying, this decrepit cottonwood becomes a tree of life. Two shiny, mustard-colored Goldsmith beetles are locked together, legs hooked to roll themselves up in leaves. As I sit cooking, giant red ants head up and down the stained, furrowed bark of a huge fallen limb, its blanched wood veined by borer beetles. Toward dusk a bizarre-looking dobsonfly flutters awkwardly from one limb to another before falling to the sand. This poor guy's wings are broken so he drags himself pathetically along. Male dobsonflies exhibit large, crossed jaws that look fierce but serve only to hold the female during mating.

The river roars by, picking up speed. I check the knots on both ends of our mooring rope before crawling into our bag. When you've only got one boat, you can't risk a breakaway.

Our third day begins with a serious tie-down of everything: the waterproof bags, the canteens and water jugs, even the tarp covering our gear in the back. Anticipation rising in our throats, we pull in to scout Steer Ridge Rapid. This is the one that capsized Major Powell's boat, the *Emma Dean*. After glassing the rapid, we plan a way through the turbulence. The problem with scouting, whether from the bank or from the "pond" above a rapid, is that it gives you a distant picture that may not reflect the hydraulics, the dynamic reality of the river.

We enter the huge tongue and shoot right into the roughening waves. One breaker slaps Barbara in the face;

another pours over the side, leaving two inches of water as the river slows. Unsettled, I look back to see nothing but frothing chocolate milk—no telltale rises that signify submerged rocks or smooth places that mark a hole. A straight shot that proved easier than it looked.

But this is no place to get caught looking back. Before we can bail, or even wipe off our sunglasses, we're daunted by another guttural roar. Around the corner looms the well-named Surprise Rapid. It's too late to scout it, so I backpaddle to buy time for a look. Only airborne globules of froth are visible beneath the curvature. So far as we can see, no obstructions mine the main channel—only large freestanding waves, well spaced in a row. Looks good from here.

We plunge straight into this turbulence. Oddly, though, these are not typical stationary river waves—they chop and hit from all sides. A curved wall tosses our bow skyward. After Barbara punches the front tube, I see her rise above me, fists gripping the rope. As the bow plunges into the trough, we're both engulfed by spray. In a twelve-and-a-half-foot boat, five-foot curlers give you a rollicking good ride. But soon enough the rapids subside as we float down to a spot we're longing to explore.

Rock Creek Ranch is both idyllic and remote. Its setting is spectacular, for here Desolation Canyon is as deep, though not as steep walled, as Grand Canyon at Phantom Ranch. The Tavaputs Plateau—really rounded, conifer-covered mountains—rises more than five thousand feet above the river. The three sandstone buildings are the first signs of human settlement that we've seen in two days and forty-two river miles. The Seamounton brothers, who settled this ranch around the turn of the century, had to pack in everything on fifty miles of rough trails. Nothing arrived by boat because, not surprisingly, this stretch of river was considered virtually unnavigable. What were the payoffs for all the extra effort required to ranch here? Perhaps the old forge speaks of a frontier desire, no doubt heightened by remoteness, to become as independent as possible. Perhaps, too, the Seamountons were living out the Western myth of creating a personal paradise in a place offering profound isolation.

The cut sandstone buildings are reliquaries from the past. A shed built with hand-squared timbers still houses rough-hewn furniture, while all around languish old wagons, plows, harrows, and other implements. Beside a ditch once used to irrigate, mulberry trees still bear luscious fruit. Hands and lips stained purple, Barbara and I saunter through the prickly cheatgrass whose tan expanses resemble a river backwater.

An import from the Steppes of Eurasia, cheatgrass spread rapidly in the West with the advent of massive grazing in the late 1800s. Although ranchers initially welcomed this hardy invader because its young leaves made good winter feed, they soon found that, overall, it offers forage far inferior to the native grasses it replaced—and that its barbed seeds stick in the mouths, ears, and feet of livestock.

After a quarter mile we pass the old corrals and head up Rock Creek. Rainbow trout flash as songbirds feather the air. Western kingbirds are nesting in a dead cottonwood where the male is flapping his wings just above his mate. With fresh beef, fruit, and milk, plus fish, birds and game galore, this must have been a fabulous place to live.

On the river we encounter a spirited group from Salt Lake as they're lunching on a delectable spread of croissants, cold turkey, and beer. We chew our carrots while someone tells a story about the exotic pools near the San Juan River. One guy figured that rafters must lose their sunglasses in these pools, so he dived for sunken treasure. While he wasn't looking, somebody tossed in the diver's own expensive shades. When he "discovered" his own pair, he surfaced ecstatically, thrusting them high in triumph. We all howl, including the guy who got fooled.

At Chandler Canyon large cottonwoods line the clear stream, sagebrush grays the alluvial flats, junipers beard the base of the cliffs, and conifers climb the higher slopes. This area offers inscriptions by the early French trapper D. Julien and an amazing Fremont Indian hideout. A hundred yards up Chandler Canyon, just above the chimney of an early cabin, the Fremont site sits among jagged slabs that have fallen, leaving hollow space and opening in different directions. This shelter may have served as a secret observation post, since one can see both up and down canyon. The Indians might have intended these slits for shooting arrows or for escaping out the back. Once accustomed to the dim light, I stare at an oval mud-and-stone granary designed to provide food for people who took secret shelter here. My finger presses into a print left in the once-soft clay to make a hand-to-hand connection, a gesture of shared humanity.

This hideout must have offered as secure a place as any. Its only secrecy problem might have been that its location could have given it away. It's situated at the injunction of canyons with streams, which is right where enemies would have looked for rock art. On this stretch of river, petroglyphs appear at the junctions of the Green with Rock House Wash, Jack Creek, Florence Creek, and elsewhere, which suggests that they may have served as signposts, territorial markings, or bulletin boards.

We've no sooner settled down at the campsite than three red rafts come heading our way. Disembarking from his flagship, a beer-bellied guide waddles into our camp to announce that his group has come all this way in just a day and a half.

"You're really pushin,'" Barbara comments.

"Well, gotta be—I'm doing a commercial trip on a private permit, so I gotta get these customers through." Rather than take his entourage around our campsite, he leads them right through it. I manage to nod weakly as the poor customers trudge by.

After supper we pick our way through the lush reeds and sedges along the bank. Kayakers are playing like dolphins in the whitewater, running Chandler Falls again and again. Across the river, tight against a salmon-colored cutbank, young cottonwoods glow a phosphorescent chartreuse. Far above, at the head of the drainage, looms an arch spotted on Major Powell's second expedition in 1871.

Orange rays extend as spokes into a royal blue sky, evoking the wheel of the fabled winged chariot crossing the sky. This image is obviously a carryover from a long-discarded world view, but so too is the concept of a "sunset." Why don't we say, "What a beautiful earthturn?" It's instructive to consider the persistence of paradigms, however outdated, that still serve some human need. It's a comfort to believe that the earth is the hub of the universe, or that we're God's special creation. While the scientifically minded may smile at such notions, today many of our most progressive minds question the established scientific paradigm as well.

As we launch early the next morning, a doe and her fawn stand among the willows, moving only their jaws while we skirt the bank. A pair of Canada geese with young are feeding among the reeds. Rather than fly off as they ordinarily do, the adults hunker down to water level as their goslings become light brown flecks of flotsam. Moments later two other Canadas ride regally through the rapid, heads held high above the froth.

After Rock Creek, McPherson Ranch isn't much. Founded in the 1880s, it too had irrigated fields and orchards, a hewn stone smithy, a red rock chicken coop, and the rest. Around the turn of the century, Jim McPherson extended his celebrated hospitality to Butch Cassidy and the Wild Bunch, who crossed the Green here on their way from Utah to Wyoming. McPherson remarked that he got along better with the outlaws than with the lawmen who chased them.

Disappointed by this stop, we're eager to return to the river where Three Fords Rapid soon challenges our skill. As the roar gets louder I stand to scout, then backwater for a time, wavering back and forth, reading the water. Finally we descend a steep chute. Currents suck us straight toward a large boulder. Failing to catch as we crest a wave, my oars fan in the foam and the current sweeps us toward the outside wall. Avoiding it will require all the strength I can muster. My feet brace against the frame and my tailbone jams against the seat. Power strokes timed with the troughs finally pull us off the wall.

"Big rock," yells Barbara. Instinctively I spin our stern 45 degrees to take advantage of the swift current around the slimy boulder. We clear it, and the treacherous hole behind it, by only a few feet.

Barbara whoops, then grabs the bail bucket. After all this agitation, the river boils and bubbles for another quarter mile, spilling sideways into the eddies along its banks. All exaltation aside, the fact is that we've been lucky. Our scouting, such as it was, didn't reveal these hazards at all. Running rapids involves spotting snags well before they approach, setting up to avoid them, and rowing furiously around the ones you didn't see.

Next comes Coal Creek Rapid, the biggest water in Gray Canyon. This time we stop to scout, eying the mammoth boulder at the top and the impressive five-foot waves. A snake churns my insides. We'd better double check our knots and carabiners and wait for other boaters. Flipping is one thing, but watching your boat drift away is another.

Much to our relief, the Salt Lake gang appears. After debating who will run Coal Creek Rapid and who will line around it, the kayakers agree to hover in the eddy until we come through, one way or another. Supported by this safety net, we follow the confident kayakers as they disappear into the turbulence, slice though the big waves, and cut across the current. Their route doesn't seem right for us, but it's too late to set up another way.

Slapping us from both sides, the cross chop rips an oar from my grip. Unpredictable waves pitch the boat and thrust us into the air. There's no point in trying to punch tubes or time strokes. Feet spread wide for balance, calves braced against the back tube, I stretch to see above the spray.

"Right, right!" yells Barbara. Dead ahead lurks a thunderous hole, ringed by whirling spray, that we couldn't see from upstream. Steering is futile. The oars become matchsticks in a maelstrom. Though cross currents push us right toward the mound of water, we somehow miss the big rock. Then our luck runs out. Like a cork spun into a whirlpool, powerful currents suck us right into the huge hole behind it. A wall of water blasts me out of my seat, onto the baggage. Everything seems to stop. Then, in slow motion, a tube rises toward the sky and Barbara springs clear, free of the rope.

Brown bubbles. Like dark beer swirling in a stein, undercurrents spin me head over heals. No point in trying to swim: I don't know which way is up. Finally my life jacket buoys me to the roiled surface where I suck in spray and choke. I rip off my sunglasses to look for Barbara. She bobs in the turbulence not far away. The *Canyon Wren* drifts downstream, upside-down.

"Swim for it," I gasp, eyes riveted on the boat, lungs finally inhaling less water than air. The river becomes a wake of boils. We swim side by side, stroke for stroke to reach the black rubber, but Barbara's hand slips off the tube. The flip lines and throw ropes are submerged, out of our reach. I grab an oar to pull myself in, then crawl up and onto the bottom. Then I extend a shaky arm to pull Barbara aboard. We sit like bedraggled royalty on a float as we approach the kayakers, who tow us toward the bank.

"You guys aren't equipped to do rolls," one jives.

Though this sounds like an excuse, there's some truth to it. A kayak can not only handle more turbulence but can maneuver much better in a rapid. Once a heavily loaded raft has entered a rapid, the same enormous forces that clear the channel of rocks oppose a rower's efforts to alter course. The river simply takes control. Still, we might have dodged that big hole if we'd seen it in time.

With six boaters giving a heave-ho together, our raft rights easily. Everything seems to have come through except for my "waterproof" camera pouch. Everyone groans as the river dribbles out of my Nikon. Barbara takes her initiation well, but I feel unsteady, shaken. There's no better way to experience the power of a river than to spin around in a hole, but it doesn't feel like a dip in a pool.

Barbara takes the helm while I recover. Though an afternoon wind wrinkles the river, she finds her strokes and stays on course. As we glide down river from Rattlesnake Rapid, I watch three red rafts pull off the river. Good. Then, panning downstream, I glass the banks for a likely campsite. Nothing but banks walled with tamarisk and gray cliffs with plants hanging below the lime line along a seep. Unseen chats whistle sharply. This stretch teems with plants and bugs and birds, but not with campsites.

We land on a beach just above Nefertiti Rapid. This site's not pristine, but it's spacious and clean even though a dirt road from Green River reaches this far. As we're unloading, however, the red armada appears around the bend, pursuing us like the Furies. The beer-bellied guide asks whether his gang of thirteen can camp here too.

"Look," I growl, "you had the site at Rattlesnake." He and his customers move on.

Above our camp an old cottonwood quivers with life. A Bullock's oriole weaves her basket, scolding any robins that come too near. She builds the nest alone, attaching long fibers, tugging through the cross fibers, and lining her basket with fuzz. The flamboyant male, his head striped like a Michigan football helmet, lands and exchanges liquid notes with his mate. His job is to give a chatter call if he spots intruders, or to flash his colors for a defensive display.

In a blur of olive and yellow, the female oriole swoops to grab a wasp from a thistle poppy. As she turns it in her mouth, its legs continue to grasp the air. She drops it on the hard sand, looks it over, pecks off a leg, and seizes it again, biting off its head. Several minutes later, after more chomping and turning, she finally swallows her prey and even picks up the severed leg. After the speed of her dive, her eating ritual seems very slow. The wasp was here one moment and gone the next, as any of us could be.

Though we savor our last night on the river, it's time to end our odyssey. Our hands are scabby, our legs scaly, and river mud cruds our hair. My thoughts creep toward such degenerate pleasures as a shower, an air-conditioned motel room, and a huge Greek salad with extra feta cheese. Once upon a time such yearnings would have led me to suggest some revelry around a fire, but I'm not in a party mood tonight. I've been humbled and need to contemplate my limitations.

The next morning finds me in a different emotional place, for I know that I'll soon experience re-entry symptoms. After a short float the deep thud, thud, thud of speakers announces civilization at Green River Beach where two sunbathers blare rockabilly from their pickup. Their poodle yaps at us as we tear down our rig. We haul our gear over blistering sand and cram everything into our van, including deposits of river silt that we'll carry for many miles.

This hauling is hard work that more boaters might have helped to share if we had traveled in a group. Yet this exertion also speaks to one of the great pluses to solo wilderness rafting: except for moments like these, there's little conflict between what one wants and what one needs to do. On this trip we've balanced a high degree of freedom with a fair amount of safety, relying on fellow travelers to help us through the rougher sledding.

But this balance doesn't come without a price. Though we've traversed a great wilderness, we've probably met too many fellow travelers to experience a full transformation into the natural world. By day three or four of real solitude, the senses attune and consciousness alters toward river, land, and sky; on this trip, such attunement has seemed more limited. In its root sense of freshening, re-creation best occurs away from groups. If, as Wallace Stegner observed, wilderness is "the place where the individual makes contact with the universe," then it's important to minimize the background noise.

Interlude

The Beavers of Escalante

Who would expect to find beavers in Sand Creek, a small tributary of the Escalante River within a complex of desert canyons? This clear perennial stream grows few trees, partly because it experiences frequent flash floods, and this desert bakes in the summer. Yet beaver dams and lodges dominate this narrow canyon.

The presence of beavers affirms both the micro-climates of the Colorado Plateau and the resourcefulness of these remarkable rodents. By felling trees, as well as by building dams, they become one of the few species besides humans to create their favored habitat. Nor are they the only ones to benefit. Standing trees killed by beavers offer nesting cavities for birds as well as food for insects, and dams made by beavers create habitat for many other species. Muskrats frequently live with beavers, feeding on their leftover twigs. And just below my boots, fish about fifteen inches long are fanning their tails to stir up silt from the bottom. Mouths flaring, they work together to clean and defend their eggs. Without beavers, fish this large couldn't survive in a stream this small.

Beavers also show foresight in solving problems. If water flows increase, beavers build small jetties above their main dam to slow the current. If this does not stem the flow, they open floodgates in their dams. When the floodwaters from Boulder Mountain roar down this narrow defile, though, these beavers have to rebuild their dams. In more open environments beavers also demonstrate foresight when they dig long canals to divert water into their ponds.

Smart as beavers are, they sometimes appear to jeopardize their own food supply. Since some trees near beaver ponds remain untouched, beavers obviously aren't dropping trees as fast as they could. Yet they gnaw completely around many old cottonwoods, killing but not felling them. In doing this they miss most of the inner bark and all of the tender branches. Perhaps in the past they've gotten away with this because their favorite food trees—cottonwoods, aspens, and willows—generate new shoots from their roots.

But why would such a smart, successful animal as the beaver exhibit seemingly counterproductive behavior? One hypothesis is that beavers carry genetic baggage, or "evolutionary drag," that's now becoming a liability. In the past the tendency to chew large trees was advantageous because killing old trees made way for more-nourishing saplings which, when left to grow, assured reproduction of the trees. Both mammals and trees benefitted. Today, however, this same behavior could contribute to the demise of cottonwoods. Because most Western rivers are no longer allowed their natural spring flooding, most cottonwood seedlings can no longer locate enough moisture to survive. In addition, cows eat or trample young trees.

Over time, then, genetic inheritance may become dysfunctional or even self-destructive. As our human ancestors evolved among faster, stronger competitors, they probably derived some biological advantage from their aggression. But today, given the deadly technologies in our hands, these same aggressive tendencies threaten our survival. While we've long crowed about our ability to reflect on ourselves, now we can apply this unique attribute to our evolutionary inheritance—our blind spots—in ways, alas, that even beavers cannot.

11.

Grand Gulch:
The Scream at Dawn

*Seen from the mountain, the country between here and the
San Juan and Colorado and beyond them is as rough and
impenetrable a territory as I have ever seen. . . . To the
east, great canyons seam the desert, cutting vermilion
gashes through the grey-green of the sage topped mesas.*
—Everett Ruess, letter from Navajo Mountain, 1934

Despite overuse after its designation as a Primitive Area in the
1970s, Grand Gulch retains both its archaeological interest
and a supernatural ambience. Like several other gorges that
drain into the San Juan River, Grand Gulch begins as a
shallow ravine on Cedar Mesa.

Kane Gulch, an old cow path, has become the main route
into this popular wilderness. As the trail enters the broad

drainage, everything screams of cows: dried splats litter the bare ground where chewed-down willows snag debris from flash floods. Nor is this degradation a new problem, for cattle have grazed these canyons since the 1880s. In 1891, after getting a tip on some abandoned, semi-wild longhorns, young Al Scorup rode well over a hundred miles from his home in Salina, Utah, to the Colorado River. After he and his horse swam across, he headed up White Canyon, the drainage that carved the natural bridges near here. Soon enough, however, well-armed Texans ran Scorup off. But Al returned for his grubstake, this time accompanied by his brother Jim. The Scorup brothers ran their cows in even wilder terrain while they hunted deer, devoured sourdough and beans, and lived in shanties or caves, less like cowboys than Indians.

Given the remoteness of this area, their hardships inevitably continued. For one thing, wild longhorns were tough to round up. After pursuing an animal at a gallop, roping it, and wrestling it to the ground, they had to hog tie it before it could get up. Then these cow punchers cut off its dangerous horns, tied its nose against a tree for a day or two, and finally led the bleeding, subdued animal to a corral. Rough work though it was, chasing "runnygades" became sport for many young cowboys. When the mythic American Cowboy sat tall in the saddle, however, he never spoke of the sawing and snubbing.

As the Scorups brought their stock across the Colorado, Al's boat almost sank when a bull tried to climb in. Abandoning the boat, Al rode the bull across the great river. "I wasn't trusting those huge waves. I couldn't swim and I knew the bull could." Once they got their cattle ashore, they encountered more problems. One wolf, "Ol' Big Foot," killed one hundred and fifty calves in a killing spree. To make matters worse, wild horses gobbled up the forage, ruining the range and forcing the cowboys to shoot hundreds of mustangs.

As herds increased in size and cattle roamed much of San Juan County, eventually the Scorups merged with the Sommerville brothers. The resulting SS Cattle Company grew to ten thousand head that grazed nearly two million acres—mostly on public lands such as Grand Gulch, which the company reserved for its best steers. By 1940 the Scorup/Sommerville operation was one of the largest in the West.

Since these canyons are too hot and dry for livestock in the summer months, each May these cattlemen drive their cows and sheep up to National Forest lands. Though barely profitable in terrain as sparse as this, grazing persists today because federal agencies charge less than ranchers would pay to run livestock on private lands.

As I step over a broken-down fence, mountain bluebirds flash their exquisite azure plumage to lift my spirits from the hoof-pocked mud. Rounded like the backs of elephants, the cap rock here has weathered gray where rainwater has leached its iron. The creamy-tan Cedar Mesa sandstone is veined with thin beds of gray limestone; since its sand is fine-grained and well-cemented, it resists cracking and flaking. All this roundness occasions a sense of release from the straight and perpendicular lines of modern civilization.

After a mile I'm finally beyond the cows: the trail is no longer pitted and the plants are flourishing again. Here, at about six thousand feet, Rocky Mountain species tuck themselves into the upper ends of the canyons. Where the stalks of green gentians reach nearly head high, their flowers host large ants and giant bumblebees. Small aspens rise from crevices, their greenish-tan trunks vivid against the dark-stained rock.

While a slight breeze animates their upper heart-shaped leaves, their lower foliage droops idly.

Along the trail bees hum on the yellow-blossomed Rocky Mountain beeplant, a roadside clover imported from the steppes of Siberia. Here it grows four feet tall and nearly covers the trail. As I stride through the fragrant beeplant, hundreds of tiny azure butterflies dance to the rhythms of my gait. Moments later, hiking on undulating slickrock fosters the illusion that I'm the first to trod this way.

Another mile down I slip under a cliff beneath red rocks that are graying in the midday glare. Just beyond the shade, an emerald-green hummingbird hangs motionless, its miniature head turning sideways to look me over, and then zooms away with a pulsing, high-pitched squeak. A canyon wren pours out liquid notes that cascade down the steps on the canyon wall. Before long this mottled-brown songster flits down to a boulder, then right into the stream where it flutters before rolling in the fine sand nearby. Apparently this dust bathing activates oil in its feathers and discourages lice. The effort that birds put into their preening suggests that their feathers must be kept in fine fettle.

Across the canyon the two-hundred-foot wall has fractured into rectangular panels, each framing a design fashioned by recent breakage, dark desert varnish, wind pocking, and water streaking. While some breaks in the sandstone often exhibit concentric lines radiating outward, older breaks develop a blue-black patina that builds over many millennia. Composed of manganese and iron oxides cemented with clays—all carried by water percolating through soft rocks— these minerals collect on smooth surfaces where bacteria and evaporation combine to form the dark polish that often reflects the blue of the overarching sky.

Anasazi rock artists often bypassed newly exposed panels in favor of ones coated with varnish that they could etch to expose light rock. Near a seep, a squiggly water symbol recedes into the black glare. Such zigzag or snake glyphs may have represented the serpentine flow of water, or they may have served as a visual prayer for rain. The Anasazi may have worshipped water as a sacred substance.

At this point Kane Gulch becomes classic slickrock country. Here its walls tower three hundred feet before curving into the rim, and its floor bursts with desert bloom. Maidenhair ferns and white columbines festoon a cool drip that sparkles as it beads the darkened rock, while golden daisies and lavender penstemons grace the hot sands. Firecracker penstemons and scarlet gilia blaze in the tawny grass. Small cacti burst into waxy pink blooms, each with an insect feasting on its pollen-ladened anthers. An emerald-green fly hovers, its wings beating silently and nearly invisibly. Hover flies actually pollinate more flowers than bees, which don't range as widely.

As the trail drops into the streambed, my boots crunch the gravel. A flash flood has sheared an already-twisted juniper in half, dangling strands of bark like drying seaweed. At the top of the cutbank, blond grasses trim the vivid blue sky. A ranger asks to see my permit. As his horse waters the stream, he reminds me not to pollute the watercourses.

Around the bend Kane enters Grand Gulch, the greatest concentration of pre-Columbian ruins in Utah. Here the Anasazi found optimal conditions for their agriculture. At Junction Ruin, the largest cliff dwelling in this area, an overhang allowed the sun to warm the buildings in winter but prevented it from baking them during the summer. Enhanced by small dams designed to catch both water and soil, a semi-permanent stream insured water for irrigation.

Since other hikers are inspecting the ruin, I seek nearby Junction Spring. Water striders are floating on the small pool. For these widespread insects, surface tension is all: they break through it as nymphs, and their adult existence depends on it. Though I'd assumed that water striders rely on pontoons, now, up close, it's apparent that their legs make indentations in the water's surface. Their back four legs dent the surface while their front legs grab and hold prey, such as the fly that may have pursued me here.

Just beneath my face, on the animated pool, expanding circles slice through each other. Amid the bugs I see my reflected face, brow slightly furrowed, with shafts of sunshine beaming around my bushy hair. In nature, I muse, we're more free to be ourselves. We don't see ourselves for days on end, reducing self-consciousness, and when we finally get a glimpse, we're less likely to care about our appearance and more apt to like the person we see.

As I look up, Junction Ruin is once again deserted, everything having fallen silent except for the whine of insects and the croak of ravens. As I cut through the brush, my boot snags on something. To my amazement, I retrieve a blanched yucca-fiber snare from the sand. In more ways than one I've been snagged by a people who departed seven hundred years ago.

The ruin itself offers more connections. Scattered in the loose sand are pot shards, miniature corn cobs, squash stems, and charred sticks from ancient fires. Closer in lie compacted ashes, sticks, yucca fibers, and other refuse alternating with thin layers of sand. Do these alternating layers indicate ancient sandstorms, or perhaps temporary abandonment of the site? Such midden heaps reveal much about how these people lived. But as Native American writer Leslie Marmon Silko warns, we should not assume that such close-in refuse heaps suggest sloth: "Corn cobs and husks, the rinds and stalks and animal bones were not regarded by the ancient people as filth or garbage. The remains were merely resting at a mid-point in their journey back to dust." Since humans make a similar journey, it was surely no sacrilege to bury the dead alongside the garbage. Rather than presuming to stand apart from nature, the Anasazi apparently included themselves within the life-and-death process.

The rooms served three essential purposes: food storage, general residence, and religious ceremony. They were built with carefully-piled, handhewn rocks, reinforced with parallel sticks supporting mud masonry and secured by rodent-proof doors. Entering one of these rooms is like crawling into a small cave with bare bedrock for a back wall. From inside it's not difficult to imagine why the Anasazi spent much of their lives outdoors.

The circular kiva, the subterranean ceremonial chamber, evolved from the pithouse of earlier Basketmaker times. Named by Major John Wesley Powell after its contemporary Hopi counterpart, the kiva featured stone benches for up to eighteen men. Each kiva had a *sipapu* symbolizing the navel through which humans emerged from the earth. Possibly to evoke a sense of this emergence, this birth of the race, kivas seem designed to feel womb-like.

The Anasazi probably would have found it strange to assume, as our culture does, that heaven is a more sacred place than earth. As Silko observes, "A rock has being or spirit, although we may not understand it." This is an important cultural assumption, for when we believe that spirits inhabit plants and animals and rocks, we tend to treat them with greater reverence and respect. Conversely, when we assume that spirit and matter don't intermingle, or that only humans have souls, we tend to exploit the physical world. The consequences of such separation and objectification are all around us. Although indigenous people were not environmental saints, we can certainly learn from them.

These ruins involve more than just a backcountry museum without labels and showcases. They abound in signs of human habitation, preserved in place by the dry climate and the overhanging cliffs. This is not Mesa Verde with its artifacts removed and everything restored. Here mounds of refuse often lie right where the Anasazi left them. There are no rangers lecturing, no tourists yakking, nothing to interrupt pure wonder. Beneath these towering walls the spirit of the Anasazi quickens. Visitors often sense an Anasazi presence, which some find unsettling or even sinister.

These ruins also connect our well-fed lives with those of people whose existences depended on the timing of rain and the luck of the hunt. If these glyphs and stones don't speak, they do spin our minds back through the centuries. On a large boulder are deep grooves where rounded stones once ground corn. The foot-long mano and metate stones lying nearby were probably used for the initial crushing and grinding. I can almost see wrinkle-faced women working the maize while they watch the children. The younger women toil down below, cultivating a patch of beans where turkeys nab grasshoppers. A young mother and a grandmother emerge from a doorway to introduce a baby to the sun. What was it like to live here, day after day, year after year, and possibly to die without seeing much beyond these canyon rims?

For full impact, such wondrous touchstones into the past belong on site, where they were left. The yucca-strand necklace, the squash stems lying about, the small cobs of maize in my palm would not impart much sense of connection if they weren't here to discover. By leaving artifacts where they are, other visitors have enhanced my experience with the people who once lived here. Hikers who encounter artifacts used to place them on rocks, which seems admirable, but archaeologists deplore such well-intentioned gathering because it removes the items from their original positions.

Worse than that, at some point plundering became a weekend job for many locals in southeast Utah. In recent decades looters have desecrated more than 60 percent of the sites in this area. In reaction, federal agents carried out a heavy-handed raid that even nabbed some San Juan County officials. The state has prosecuted other diggers. Of course these crackdowns have antagonized locals who resent any infringements on their economic options.

Some of the worst damage, however, was done much earlier. After the Anasazi departed around 1300 A.D., these ruins stood largely unvisited for centuries. Once white settlers arrived, they and their livestock did considerable damage. Early explorer/archaeologist (or, as some would say, pothunter and grave robber) Richard Wetherill was chasing cows when he stumbled on Mesa Verde, a masterpiece of Anasazi architecture.

Just after his startling discovery southeast of here, he led the first expedition into Grand Gulch. It must have been thrilling to find the dwellings much as they'd been left, many centuries before. In 1893 Wetherill dashed off an excited letter to announce that the remains of a much earlier Basketmaker culture underlay the ruins and debris of the Cliff Dwellers. Unfortunately Wetherill and others carted off the artifacts, typically selling them to Eastern museums. While these diggers slavishly followed the "Western spirit" of exploration and exploitation, they also diminished the scientific potential of many sites. By 1905, when Congress outlawed such practices, they had stripped Grand Gulch of its major treasures.

In the late 1980s the Edge of the Cedars Museum outside Blanding began to restore a lost heritage. Volunteer "reverse anthropologists" scoured these canyon walls for the names of early excavators and also combed their field journals to determine what was found where. The goal is to retrieve the relics, housing them in Blanding, Utah, as close as possible to where they were found. Although these researchers initially encountered difficulties accessing private and public collections, they now have success stories to tell.

In the rock art above me, timeless dancers and ancestral gods play on the mind. The gods become visible in the masks, the dancers invisible behind the masks. One dancer is playing a flute, head tilted and hips bent in the classic position of the celebrant. Like the revelers in Keats's famous "Ode to a Grecian Urn," this figure is frozen on a note that echoes over time. He is a *Kokopelli*, the phallic but hunchbacked flute player associated with cunning, fertility, and hunting.

To Hopi descendants of the Anasazi, he still pipes as a kachina or deified ancestor. Like their notorious snake dances, the Hopi ceremonies that celebrated his presence were baldly sexual—and powerfully primal. These "obscene" rites offended early white observers. Some turned away to smirk, while others, such as early Mormon missionaries, were thoroughly shocked. Such ties to instinct, to the earth, and to nature, sanctify deeply religious connections that Anglo-American culture has found less sacred than profane.

Suddenly the ancient pictures dim like shadows. For no

apparent reason my throat convulses, leaving me short of breath. More eerie energy creeps in, so I leave.

The trail leaves the stream, cutting through horsetails, willows, and tamarisk. I scramble up the loose sand of the bank, entering a sylvan glade in the desert. Bow-trunked scrub oaks, usually growing in clumps, shoot up about fifteen feet. Leaves on the huge cottonwoods click in the breeze. Beneath them, sunlight and shade dapple the sand, evoking a sense of floating in light like a park scene by Pierre-Auguste Renoir. As a gust wanes here, it soon rattles the cottonwoods downcanyon. These gnarled, venerable Fremont cotton-woods rank among the glories of Grand Gulch.

Cottonwoods are a druid's delight throughout the year. In early spring sticky, almond-sized buds burgeon into drooping maroon catkins. From dawn to dusk, trees in bloom often host bees audible fully a hundred feet away. In April, as orioles begin to perch and sing, heart-shaped leaves burst forth in a delicate chartreuse/orange. By late May the seeds crack open, filling the canyon air with cotton puffs, while orioles weave their baskets in the gnarled twigs. All summer the dark green leaves contrast vividly with the red rocks, white clouds, and blue skies. In August swollen streams pile debris around exposed roots.

In the fall the cottonwood's golden foliage, like rocks that hold summer heat, radiates warmth into the cooling air. By November their golden spangles spiral stem-first to the earth, where they fashion a mosaic beneath the spreading branches. In winter bees in the trunk fan their wings to stay alive. As snowflakes descend, these gray boughs and stubby twigs reach toward the gray-streaked canyon walls, interlacing plant and rock, earth and sky.

Unfortunately old-growth cottonwoods are not faring

well. With only about twenty or so significant stands left in the Southwest, the cottonwood and willow community has become the rarest forest type in the U.S. Cottonwood seeds germinate quickly on the moist alluvium left by stream run-offs, then send down roots that must reach water in a short time. The tamarisk, their chief competitor, also germinates on moist floodplains but, unlike the cottonwood, sows seeds all summer. Moreover, tamarisk seeds root with less moisture and sink taproots deeper and faster. As drought-resistant tamarisks enlarge their range, cottonwoods are dying out, especially as irrigation removes more water from drainages and dams check spring flooding. This decline is receiving some attention from the Bureau of Land Management, which is starting to protect saplings from cows and off-road vehicles.

After tromping through oak thickets and cottonwood groves, I pass Todie Gulch. Here the canyon wall divides into trapezoidal towers rounded by eons of wind, water, and blowing sand. Farther down Grand Gulch, I camp near bands of gray-green sage, of yellow beeplant, and of lavender-brown cheatgrass. Dwarf forget-me-nots waver just four inches above the sand, their minute blooms like turquoise beads in a Navajo sand painting. In the twenty years that this canyon bottom has not been grazed, the vegetation has made a remarkable comeback.

As the waning light creeps up the canyon wall, it casts a peachy glow on the darker side of the canyon. Two large hawks glide into their nest hundreds of feet above me. Song-birds burst into a final crescendo, perhaps "whistling in the dark" to allay their fear of owls. Grainy dusk begins to fall, yielding those mid-spectrum blues that we primates see better than, say, hummingbirds that are more attuned to reds and yellows. Humans can see the coming of darkness that many other animals only feel. Nevertheless, they know. From inde-cipherable directions birds pipe single notes as they close the soft-dying day.

Tucked into the sandstone wall above us, canyon tree frogs begin a croaking duel that keeps me awake. Never, though, have I felt happier to hear frogs. For many years grazing bottomlands had endangered the relict leopard frog in the Southwest. By the late 1980s, however, amphibians had suffered a sudden, widespread decline that signaled possible ecological breakdowns. No one theory explains the demise of all these frogs, toads, and salamanders, though local drought, habitat loss, global warming, ozone depletion, and acid rain are suspected contributing causes.

Matted by the black canyon walls, a band of stars is shining. The heavens also impressed the Anasazi, for their star motifs appear beneath overhangs or on the roofs of "star chamber" caves. Studies of the "archeo-astronomy" of Indian cultures have established the significance they attributed to massive explosions or supernovas. When the spectacular Crab Nebula first appeared in 1054, it shone five times brighter than any other star in the sky. To most ancient peoples, astronomical displays evoked the mysteries of the universe and indicated the seasonal cycles so crucial to hunting, gath-ering, and farming. To us, alas, the stars and planets less often inspire a sense of the rhythms of the earth, moon, and sun.

Remarkably, since Grand Gulch has become so popular, I've seen no one since Junction Ruin. The sense of isolation is strong, intensified by the barrenness of the canyon walls that evokes a liberating blankness of the rational mind. I'm enjoy-ing the free play of my consciousness even though it's costing me sleep. The ancient ones, I suspect, slept better beneath their turkey feathers, lice and all, than we do in our fancy sleeping bags. Despite the fatigue, however, I sleep fitfully. After a dream about getting buried alive in a kiva when the roof mysteriously falls in, I awake in a cold sweat. In a minute or so I hear a rustling of leaves not far away. A stick cracks. My heart races. "Probably just a deer," I tell myself as my stomach gurgles. I peer out. A half moon as yellow as a cougar's squinted eye stares down on my little tent. At dawn I look around but find no tracks other than my own. Strange. Those footfalls seemed so close. Spooky.

I think about other people who have reported strange events in the Grand Gulch area. Writer Rob Schultheis reports disquieting experiences among archaeologists, who are usually unreceptive to psychic or supernatural events. According to one tough-minded woman completing her Ph.D. in paleoecology, she ambled toward the Anasazi graves one evening, looking for her watch. Down in the graves, feeling around for her watch, she heard something:

> "I don't know why, but I suddenly felt afraid," she told me. "I ducked and crouched at the bottom of the pit. As the person, or whatever it was, approached, I heard it singing and chanting in a woman's voice, in a language I'd never heard before. I saw a figure pass in silhouette above me, still chanting in low tones that didn't sound human."

Fearing spirits, even today many Utes and Navajos will not approach Anasazi sites.

The next morning I set off downcanyon. Tangled roots protrude from the sand, still holding pieces of stringy bark they snagged during a flash flood. Tall blond grasses catch the early light while the sandbank below remains in deep shadow. As the bank drops to stream level, a wild prairie rose shoots from a crack in a boulder. Its buds are much redder than its deep pink blooms, its aroma more racy than an old-fashioned climber. Around each bend cliff houses sit like mansions on a hill.

Throughout its fifty-six-mile length Grand Gulch hosts varied wildlife and grows luxuriant greenery, but its cliff dwellings impart its special ambience. After passing several impressive structures on the right, I scramble up to Wetherill Ruin. Attracted to this site because it was a storage and burial area, Richard Wetherill inscribed his name and the date when he dug here. Unfortunately, its most important artifacts have since lost any value to science because of inadequate labeling.

In Sheik's Canyon, named for the turbaned mummy found here, Wetherill also exhumed "The Princess," a woman covered by baskets and a turkey-feather shroud embellished with bluebird plumage. Her body was painted yellow, her face red. The nearby gallery of petroglyphs and pictographs here casts a spell. Here ancient rock artists created serpents, birds, bighorns, and anthropomorphs, those square-shouldered men with chests like shields. There are also unusual figures such as a woman in the throes of breach birth and, most remarkably, a free-floating yellow head wearing a reddish headdress and a triple-banded Green Mask.

Somehow, possibly by drawing on the inscrutable powers of the rock, the eyes behind Green Mask stare me down. I want to meet the apparition behind the mask, the spirit within the rock, but somehow I can't retain my stare—or even lift my eyes. Sinister. I stand riveted, open to a primal connection far beyond intellectual speculation. My voice no longer sounds like my own. I seem to be hearing myself from a distance. I've never felt so superstitious in broad daylight before. I stumble off, humbled and silent, wanting to believe that I'm woozy from the heat.

Sunlight trickles through the scrub oak, puddling into a plunge pool. Though only six inches deep, this pool offers relief. I plop down without apology to the tadpoles, my sense of selfhood reaffirmed by the size of my splash. As my head tilts back into the rippling stream, my hair floats like waving weeds. Behind them, the sky pulses like a blue flame. Far above, the rocks wriggle as energy of some sort wavers down the curved canyon walls.

Still unsettled, I slog through the stream oblivious to the sun and heat. Back at camp, I sip wine to quiet mind and body. Pinyon jays squawk from the scrub oaks as rufous-sided towhees rustle the dried leaves, rasping like catbirds. After a spaghetti supper enhanced by a split of chianti, I scrub a saucey pan with local horsetails.

Just above me, mourning doves gather in a scrub oak. Their pale plumage with a lavender wash on the shoulder brightens in the twilight. One bird puffs its throat—probably a male in need of a mate—and wails the plaintive coo that gives these birds their common name. Others drop down, jerking along, heads near the ground in search of seeds. Such flocking offers birds advantages. If we assume that a predator will often strike the nearest prey, staying together helps to reduce the domain of danger for most of the birds. Moreover, as a member of a flock, an individual can reduce the time it spends watching for predators and increase the time it devotes to locating food. These benefits apply especially to ground-feeding birds that frequent habitats where both cover and food are scarce.

Before dark I dig a cathole and squat down to business. Right in my face a purple-and-white-striped penstemon glows in the lavender twilight. Since I'm so close and have plenty of time, I look closely at this desert flower. Its puffed blooms balloon into poliwogs with gaping mouths, protruding lips, and fuzzy tongues. Penstemon means five stamens, but a closer look inside reveals only four. The fifth, I recall, has lost its anther and evolved hairs, which explains why many penstemons are called beardtongues. As a genus, penstemons are well adapted to the desert. Over twenty species appear on the rare plant list in Utah alone.

Since excretion is basic to all life, it affirms our kinship to other living creatures. But this requires that we unload some cultural baggage. We Americans are, after all, the great grandchildren of the Puritans who saw bodily functions as evil. Early in the 1700s, for instance, the famous Puritan leader Cotton Mather was "making water at the wall" when a stray dog trotted up and lifted his leg. Mather, the Divine who shunned his own "lower nature," shuddered in disgust at this link. Little did Mather realize, as Zorba the Greek clearly did, that he may have held a key to sacred kinship right in his hand.

Bats dive just above my head as the Milky Way powders the sky. By candlelight I scribble in my journal about my experiences with the rock art and especially with the Green Mask—not the best subjects to explore before bedtime. If these were surges from my unconscious, they may well suggest that I've been doing all too effective a job of living in the modern world and repressing the mysterious, the irrational, in order to maintain my self-image as a good, reasonable person. To grow, I'll need to own and embrace my darker demons.

Just before dawn the loudest of alarms jolts me into consciousness. Before I can sit up, the unmistakable scream of a mountain lion pierces my tent again. Never have my eyes sprung open so fast, never has my heart beat so much like a drum in my throat. A third metallic scream—something like the stripping of gears—comes from just outside. Fumbling in haste I unzip the window. Nose to the mesh, I peer into the dusk. Nothing but brush. Then the falsetto caterwaul comes once again, followed by guttural growls lasting several seconds. With only brief lulls, these screams and growls continue for several minutes.

Finally, just as it's getting light enough to see, the last scream and growl echo down the canyon. A long tawny tail disappears into the dusky willows. This was probably a mother cougar with cubs who doesn't want me around. Like beavers, cougars are a keystone species, one whose presence directly or indirectly affects many other organisms. Here, for instance, cougars probably make it too dangerous for deer to forage at certain times or places.

Cougars favor these canyons because mule deer, their main prey, congregate in the lush bottoms rather than up on the exposed mesas. Deer need cover and shade, since they start to overheat above seventy degrees. In the bottomlands, however, they run a risk since there's more cover for cougars and there's less space to run. Thus these verdant canyon bottoms provide tight niches for wildlife and it's natural for cougars to defend turf. On the topic of close encounters with large predators, Ed Abbey quipped that "If people persist in trespassing upon the grizzlies' territory, we must accept the fact that the grizzlies, from time to time, will harvest a few trespassers."

It's important to know that mountain lions prowl these canyons. We become more alert, more alive when we feel stalked or hunted. In a healthy way it's humbling for us to step down from the top of the food chain. Wild animals also connect us to a different consciousness, joining us to our evolutionary past. As Jungian storyteller Clarissa Pinkola Estés observes, "No matter where we are, the shadow that trots behind us is definitely four-footed."

I pack quickly, hardly eating breakfast, before the sun soars above the canyon rim and streams on the leafy path. The cougar made her point and I got the message.

Along the stream caterpillars feast on ash trees that glow a hot chartreuse in the early morning light. Cool, moist air and dewy willows brace my arms and legs. With each soft footfall, each squeak on the moist grass, the ground resounds with the chants of the Anasazi. My mind's eye fills in the dogs and turkeys running about as ruddy-brown people emerge from the small doors.

As the water striders watch, I fill my canteens at Junction Spring before starting up Kane Gulch toward the ranger station. This morning I'm seeing more of Kane not only because I'm hiking slower but also because, heading up, my eyes aren't as busy scanning the terrain underfoot. About halfway the wavy, gently sloping sandstone is particularly alluring with its marbled layers of peach and strawberry on a deep cinnamon background.

This is a splendid day, with flat-bottomed towering cumuli sailing like clipper ships on a vivid blue sea. Except for the weight of my pack, no bodily discomforts interfere with total sensation. Wild roses garland the trail, big sage spices the air, and green gentians sway in the breeze. Mountain bluebirds cavort just above the rim, looking for grasshoppers and then soaring toward the clouds. I want to reach out toward everything.

As walls drop down and the canyon widens, I emerge from Grand Gulch feeling as though I've surfaced from several days in an underworld of light and dark spirits. Scarcely hesitating long enough to unsaddle myself, I head for the water cooler and the air conditioning in the BLM trailer. The ranger, an eco-conscious sort, recalls when several locals pulled his ponytail and offered him a free haircut.

Back in Blanding I talk with some wranglers about mountain lions. One ruddy fellow squints his eyes as he recounts how he trees the beasts with hounds before shooting them and watching his dogs tear them apart. Underlying this man's resentment of cougars is his realization that in marginal ranches, losses to predators make a significant difference. Nature remains the adversary for most ranchers. To understand rural attitudes, one must consider their struggles against the raw forces of nature. As the Scorup brothers learned, longhorns are ornery and remote canyons make them damn hard to locate. And over decades of eking a livelihood from a high desert, most ranchers have come to view the natural world in very utilitarian terms. Whereas they descend from settlers who sought to tame a wilderness and make a living, we recreationists seek wilderness in order to escape civilization and to learn more about ourselves.

After days and nights of bad vibes, strange sounds, weird stare-downs, and cougar screams, I too understand that wilderness is not necessarily about peace of mind. Grand Gulch remains wild, mysterious, sometimes disquieting. It's one thing to wonder about the primal, or to yearn for more of it, and quite another to actually experience it. It's one thing to understand that wilderness can drive us out of our over-civilized heads, into a serpentine dance of the psyche, and another thing to process the experience.

The Battle for the Bureau

Somewhere in the Southwest a park ranger mentioned to a rancher that before the arrival of livestock, elk and antelope grazed on grasses that grew stirrup high. Hairs on a leathery neck bristled. When the rancher called her a "radical environmentalist," she was astounded. After all, she'd never used the "O" word. We don't talk Overgrazin' in these parts, pardner.

The inflamed environmental politics of the West are fueled on the one side by the still-smoldering Sagebrush Rebellion (reincarnated as the "Wise Use" movement) against federal agencies perceived as "blocking development," and on the other by a passionate commitment to preserve some of the most magnificent country on this planet. Since the 1970s government managers have shown increasing concern with the condition of public lands, but the resulting regulations have ignited deep resentments, especially among ranchers. Their rallying cry was simple: Don't accept restrictions; seize the public lands and bring them under local control.

Fueled by decades of resentment toward "the Feds," the rebellion targeted the environmental movement, which it associated with "politicians and bureaucrats." Utah senator Orrin Hatch, who exulted about what he called "the second American Revolution," regarded everybody concerned about range ecology as "toadstool and dandelion worshippers." In contrast, ranchers saw themselves as patriotic rebels against Big Government, standing tall as they rode into the sunset of rugged individualism. "Wyoming Is What America Used to Be" claimed a sticker for a state that brands a bronc rider on its license plates.

For years rural interests have battled for control of the BLM lands near Moab, Utah. On the 4th of July 1980 the Grand County commissioners hung an American flag on a bulldozer that gouged a new road into a roadless area to disqualify it for official Wilderness designation. The environmental community registered righteous outrage, but the BLM chose to ignore this federal offense.

Nor was this an isolated instance, for the BLM often failed to enforce its own regulations and county commissioners regularly overrode federal law. Around 1970 the BLM recommended protection for Mancos Mesa, a large, isolated area known for its undisturbed flora and fauna. The agency failed to enforce its closure recommendations, however, when it allowed Gulf Minerals to blast roads for uranium exploration. By 1990, when the BLM attempted to restore the area by reclaiming these illegal roads, the San Juan County commissioners tried to block reclamation. Fortunately, the Board of Land Appeals ruled against local attempts to set policy on public lands.

Though the Sagebrush Rebellion lost its initial momentum with the demise of James Watt and other sympathizers in the Reagan administration, disaffected ranchers soon began to find supporters among other land users. Off-road vehicle groups, for example, beefed up their political muscle and joined the "Multiple-Use Coalition." When environmentalists pointed out that wilderness law already allows for multiple use, the group re-christened itself the "Wise-Use Movement." The Wise-Use guys found an ally in a BLM district manager who had already clarified his position: "Any time an environmentalist says I'm opposed to it, I'll always take the opposite point of view."

In the ensuing years, off-road vehicles became a focus in the conflict. The Moab Jeep Safari began in 1966 with a handful of jeepers riding on existing rough roads. By 1990, however, the event attracted fifteen hundred off-road vehicles grinding over twenty-two trails, many of them traversing Wilderness Study Areas. Anyone familiar with off-roaders knows that no route, however exciting, satisfies them for long: they want to make more tracks for new thrills.

When the Moab District managers agreed to allow hundreds of four-wheelers to grind into a Wilderness Study Area, the Southern Utah Wilderness Alliance (SUWA) led the fight for a reversal. SUWA's Michael Heyrend watched them churn up Arch Canyon near Grand Gulch: "It sounded like Patton's army," he recalled. BLM officials denied that they had failed to enforce their regulations, but soon afterward the Moab office received a new manager.

Citing many abuses, environmentalists took the BLM to court. Joe Feller, law professor at Arizona State University and author of *How Not To Be Cowed*, challenged the huge grazing allotment that includes Comb Wash in southeastern Utah. Working with National Wildlife Foundation lawyers, Feller contended that the BLM could not issue a grazing permit for an allotment until it had established exactly what damage is likely to occur. Once cleared of cows, Feller demonstrated, the Grand Gulch Primitive Area had restored itself significantly in just two decades: plants once again held the sand and soil. Government Accounting Office reports showed similar improvement in many other stream-bottom areas throughout the West. In a 1993 decision a federal judge ruled that canyon bottoms fare better without cows and that BLM malpractices have destroyed archaeological treasures and degraded recreational values.

While the Comb Wash area will improve and the BLM will have to reform, the battle is hardly over. Corporate-funded groups such as People for the West continue to work with county officials to block preservation and promote development on public lands. Citing section 2477 of the archaic Mining Act of 1866, local-control interests claim that over four thousand obscure tracks in five Utah counties qualify as guaranteed rights of way into federal lands. If such claims are allowed to stand, any cow path could not only preclude wilderness designation but could also become a new road into unspoiled country.

Federal land agencies will continue to operate under intense pressures from mining, grazing, lumbering, and off-road groups on one side and from environmental, taxpayer, and consumer groups on the other. But no longer is the Utah bureau the "captive agency" that Secretary of the Interior Stuart Udall encountered in the 1960s. Today the "good ol' boys" who filed away inconvenient agency guidelines are slowly heading out to pasture.

Though the BLM may no longer be the primary focus, similar clashes continue well into the 1990s. Conflicts between pro-development locals and pro-preservation newcomers have erupted far beyond Moab, in otherwise quiet Utah towns such as Springdale and Boulder.

12.

Blissed and Blasted
on the San Juan

*To start a trip at Mexican Hat, Utah, is to start off into
empty space from the end of the world. The space that
surrounds Mexican Hat is filled only with what the natives
describe as "a lot of rocks, a lot of sand, more rocks, more
sand, and wind enough to blow it away."*

—Wallace Stegner, *The Sound of Mountain Water*

My head swirls like the eddy at my feet as Barbara trots down
the ramp, looks all about, and gives the heave-ho. Once
we're underway, ripples fret on our bow.

"We've gotta be forgetting something," Barbara mutters,
scanning the launching area one last time. I secretly hope we
are leaving something behind—like the old tapes that melt on
the dash.

The tires of a big RV growl across the Mexican Hat bridge. While the San Juan Inn glides by and the German tourists wave *auf Wiedersehen*, we enter fifty-six miles of river wilderness. Moments later the canyon walls begin to expose thirty million years of the earth's history. At Mendenhall Loop the San Juan has nearly gnawed through the wall to shorten its channel. As the river wriggles into its seven famous Goose-necks, it reveals one of the world's great entrenched meanders. When the entire central massif of the continent rose to current levels, the ancestral river had already established its meander-ing course a mile above where it now flows. Over geologic time the San Juan found itself, in the words of geologists Don Baars and Gene Stevenson, "trapped in its own channel with nothing to do but cut downward." Pebble by pebble, grain by grain, this gorge has widened and deepened.

The canyon walls rise in steps to over a thousand feet, their rims capped by a hard layer of Cedar Mesa sandstone that breaks into sawteeth. The cliffs glare in the sun, picking up grays from the clouds much as the rose-bottomed clouds reflect reds from the rocks. Clusters of fossil coral, brachi-opods, and even green algae from Pennsylvanian times out-crop from the undulating layers. Upstream, oil oozes from similar bioherms where trapped organic matter, aided by intense heat and pressure, permeates the porous sedimentary rock.

Few plants grow here, and fewer bugs in turn feed fewer birds and reptiles. This naked landscape does, however, allow bird songs to carry great distances, far beyond eyeshot. Ledge by ledge, the trill of a canyon wren descends the scale. Then eloquent silence falls, broken only as the river gurgles under a snag. Twelve hundred feet above us, where the Muley Point Overlook marks civilization's last stand, a swath of sky rides between the rim rocks. We're no longer on the earth but in it, snaking through its twisted innards.

Mechanized thunder explodes the peace. Two huge B-52s streak just above rim level; the tourists up there at the overlook must be deafened by the noise. With such sleek lines, these engines of mass destruction are seductive, but their terrible beauty also conjures images of nuclear war and saturation bombings. The airplanes quickly disappear behind the rim, thundering into the convoluted canyons. The river's flow resumes, the rocks breathe again, but my heart keeps racing.

Just downstream a flotilla of Outward Bound sportyaks has beached near the Honaker Trail. Ed Abbey called these rent-a-tubs "the Tupperware Navy." It takes over two miles to climb the twelve hundred feet to the rim along a trail constructed long ago to reach mining claims. Prospecting began here in 1892, when tales of fabulously rich deposits of gold flakes lured twelve hundred prospectors. Like prospectors, pioneers often assumed that wealth awaited them in the West if only they could endure the hardships of finding it. But the gold never panned out here. American dreamers departed with little to show for their travails.

If we hike up the Honaker Trail, we'll catch the late afternoon light on one of the most famous landscapes in the West: the plateaus, buttes, and spires of Monument Valley. Fractured by the earth's folding and sculpted by wind, rain, ice, and the river's erosion, these uplifted layers weather into a grand spectacle of size, space, and time. After serving as a backdrop in hundreds of ads and movies, this Navajo land has become a national icon.

Capped by remnants of shale and conglomerate, the mesas, buttes, and pinnacles exhibit deep orange cliffs of de

Chelly sandstone rising from pedestals of soft, sloping shale. About two hundred and seventy million years ago, this was an area of wide floodplains and slow streams that provided habitat for large, awkward, lizard-like ancestors of the dinosaurs. Beyond its amazing geologic grandeur, Monument Valley is also important because, with its overgrazing by sheep, it reminds us that Indians are not the environmental saints they are sometimes imagined to be. When human populations exceed the carrying capacity of the land, both the people and the earth suffer.

The Honaker Trail is no stroll in the park. Already blistered by the glare, we wad wet towels under our hats. Because this cliff takes such a blast of sun, there's little greenery. This spring in particular the plants seem stressed; the prince's plumes can muster only a stunted golden spike. Only a fishhook cactus, its pink-purple flower looking like silk organza, looks even half-way healthy. The relentless sun bores into our shoulder blades, depleting our canteens. Though not steep, this trail is rough and mid-afternoon isn't the time to climb it.

Back on the beach two pubescent boys are fashioning a River Goddess. Covered with sand, their hands caress the torso of their voluptuous sculpture, but they freeze as we traipse by.

"It's art, you know," one squeaks cleverly. Right.

"She's got a bikini on," pipes the other. I just nod knowingly, recalling my boyhood forays into old *National Geographics* stored in the attic.

Reluctantly we relaunch our boat to look for a suitable campsite, spending an hour in flatwater shadows until we finally slip into a cove where the sandpiper and heron prints are fresh. The cleanliness of campsites today tells a success story. Once trash, fire rings, and human waste defiled these banks. By the 1960s, when dams began to limit the ability of rivers to flush themselves each spring, problems with garbage and sewage became acute enough to call for stricter regulations.

About that time, too, a new ecological consciousness became popular, encapsulated in the wilderness ethic of "take nothing but pictures, leave nothing but footprints." Today BLM regulations are strict yet river runners largely observe them. In a short time, then, we've educated ourselves and changed our behavior. Low-impact camping is only a small step, but it raises a more general question: How could I minimize my impact on the earth itself?

Unfortunately we Americans have been less effective against industrial and corporate threats, such as the 1972 oil spill that fouled the San Juan. Though few signs of that spill remain today, we can't count on comebacks because the stresses we're putting on ecosystems are so severe. Most Western rivers no longer carry enough water to absorb any additional pollution.

This backwater makes a great place to play, to venture into the river and then, after sweeping an arc, to stroke swiftly toward shore. The current is always stronger than it looks, though, and my friend John almost got in trouble doing this. He had to pull himself back against the current, rock by rock, while the San Juan nearly swept him away. So I resist the temptation to swim out to midstream, where I'd probably get swept downriver.

As we unload, we find that our river bag of clothes has soaked up water. The contents of my wallet now cover a smooth rock: my credit cards seem more plastic than ever out here and the paper money, scattered by a gust, soon becomes litter. My watch has stopped, but what's its use here? Time becomes inner and outer flow, ticked by our hearts and the river's lapping. We sip wine and savor the tans and blues bouncing on the river's ripples.

Our earlier highway hassles now drift downstream. Since I had run this shuttle before, years ago, I drove the lead vehicle. We zigzagged up the switchbacks out of the Valley of the Gods, up the old Moqui Dugway to the top of Cedar Mesa, passed Grand Gulch, and finally headed back down toward the San Juan. But there was no sign for Clay Hills takeout, so I drove on, ending up, ignominiously, at Lake Powell. As we retraced our way, we finally located the BLM sign for Clay Hills riddled with bullet holes, face down in the red dust.

Despite getting lost and covered with dust, we immediately responded to the splendid landscape. Along the river lush grasses flamed chartreuse in the back lighting. In the distance Navajo Mountain arose as a blue mound to the southwest, Bears Ears and Abajo Peak as gray bumps to the northeast. As the sun slipped behind the rounded hills, it cast a peachy glow on the rosy clouds. Somehow I'd forgotten the clarity of the air, the sense of space, the quality of the light, and the magic of a riverscape where time bubbles by.

Tonight, at our beach where the liquid light of the earthy river massages our toes, those highway hassles are gone. Daylight becomes waxy and golden as it lingers on the uppermost rocks. The sky turns a Wedgewood blue, then a slate gray. Dusk descends. Suddenly, as though spotlit, the canyon wall radiates gold and silver. The light becomes grainy, like dawn seen through sleepy eyes.

A full moon beams a silver wash on ghostly rocks. High on the canyon wall, a giant shadow of King Kong looms in the darkness of a *film noir*. It takes minutes of rapt staring to perceive that this giant shadow is really a rincon, a cavity in the canyon wall. It's true that wilderness can seem sinister, like a place where the repressed uglies in our ids could run amok. But the lighter side is that with no one around, we can become kids again, naked in the sand. Wilderness isn't just about preserving the howl of the coyote; it's about discovering the wilds within, both the howls and the giggles that society can squelch.

A breeze billows my thin cotton shirt, whisking away the gnats. The river's soft moist breath is deliciously sensuous, a feast for pores parched by the desert wind and sun. Freed from self-consciousness, excited by the moon, finally barefoot in the sand, I let my fingers dance on my bare thighs. Ah, the sweet carnal life, *au naturel*. My heart sings. How wonderful, chants my physical being, to feel released from the needs of the ego. It's scary to ponder how, in our culture, we've become so good at sensual repression. And as we've come to ignore our physical beings, it's become easier to distance ourselves from the physical world. When the moon finally bulges over the rim, I bellow with joy from way down deep.

In the morning we face the sporty stretch. From Honaker Trail to Slickhorn Gulch the San Juan drops over eight feet each mile, a steeper gradient than the Colorado River in Grand Canyon. Since the carrying capacity of a river increases as the sixth power of its speed, a stream running at two miles per hour can carry particles sixty-four times as large as one flowing at one mile per hour. As a result the San Juan is notorious for its sand waves, whitecaps that surface suddenly when the current slows and the sediment exceeds the carrying point. The San Juan hauls a heavier load of sand, silt, and clay than any other river in the U.S.

Today the river is running at about fifteen hundred cubic feet per second, low for spring runoff but good for rafting these rapids. Lower flows make rapids more "technical"—more exposed rocks require more skill. Higher water means greater speed and bigger waves but usually requires less skill because the current carries the boat right over most hazards.

Side canyons soon begin to empty in, dropping debris that often creates white water. But Ross Rapid, which roars just ahead, varies from this pattern. It's been formed by avalanches from overhanging cliffs, not by rocks from side canyons. Ross gives us an invigorating ride on its three-foot waves. Six spectacular miles below Ross, our boat glides into a crescent-shaped slip. We hope to hike up Johns Canyon but find that its floor hangs too high to reach without ropes. Just below Johns, the "old yeller" layer we've been tracking slips below river level for the last time.

According to river lore, Government Rapid was named after the government party that capsized here in 1921, but in fact the Miser/Trimble party, led by superb boatman Bert Roper, didn't encounter any problems. According to river historian Jim Aton, the mishap story may be the creation of Norman Nevills, who led trips along this stretch and liked to entertain his customers. We scan the cliffs for the landmarks—first the cowboy hat rock, then the bird rock—and pull over to scout Government Rapid.

We scramble up the cutbank, grasping for roots, then saunter across rocks and gravel. A broom snakeweed bush, white petals drooping from its long broomstraws, thrives on this hot alluvial fan. Silver spots gleaming, a cinnamon-orange fritilary butterfly lures me in for a closeup, but then flits away. Above the roar Barbara shouts, "Stop chasing butterflies!" Evidently she feels them in her stomach.

Government is tricky because the river races around a sharp bend. To run it we'll have to stay initially toward the outside to miss some snags, and then cut back inside to miss two impressive boulders near the wall. The key is to set up just far enough from the jagged wall so that centrifugal force doesn't throw us into it. Our boat clears the snags but the current sweeps us toward that wall.

"Pull, pull, pull!" Barbara bellows as my feet whiten against the frame. Spine braced against the seat, I heave all-out strokes. No time for breaths. Instinctively I spin us perpendicular to the wall, in prime position to oppose the centrifugal force.

But I've pulled one stroke too many. Just ahead a rock juts out, so I push one stroke back toward the center. Then I bury my right oar, swinging the stern to the right. We clear the rock and slide nicely down the chute. Eyes still riveted on the froth ahead, Barbara flashes the "perfect" sign.

When brown water sloshes around our ankles, however, Barbara turns around, puzzled. Did a big wave slip over the tube? The raft rows like a tanker. Despite vigorous efforts to bail, we can't keep up with the leak. Groaning into strokes, I imagine various causes for a puncture until a broken string points to the culprit: me. I had tied an ammo box to the frame with twine. When we dropped down that chute, the box broke loose and gashed the floor.

Should we camp here or lumber on to Slickhorn Gulch, three miles downstream, for a better site? Finally we round the bend above and see the break in the canyon wall. Our field glasses reveal boats pulled up near Slickhorn, so we'll have to camp upstream. The big problem, however, is gray mud that adobes the rocks and trees. Sometime last year a flash flood surged through and crested eight feet above where they are now. The river drains large areas of Mancos shale and Morrison limestone, both of which weather into clay.

Aground in muddy water, I fan my oars like a fisherman stranded at ebb tide. Barbara hops out, rope in hand, but nearly goes splat in the muck. After making a few salty comments, she prospects for flat campsites. I sink over my knees when I disembark; the gray muck belches when I finally break its suction. As I slosh toward the boat, I smear everything—cooler, frame, seat, oar grips.

After dragging our rig up the mudbank, we tear it down. It'll need a big patch, one that will require most of the rubber in our repair kit. After she props up the stern, Barbara crawls under the raft to push up while I press the patch down. As we wait for the patch to dry we sit down to drink a beer, almost dazed by the heat. Now we have to deal with the slick mudflat that lies between the boat and our campsite. We start with lighter loads, but our feet fly out from under us anyway. Charlie Chaplin couldn't have topped this. Finally we kick steps into the slippery places and drag our large waterproof bags. Running rivers isn't all moonlight in the canyons.

As their kayaks, canoes, and rafts float by, boaters scan the banks for a camp. Everybody wants to stop at Slickhorn, so we shouldn't be surprised at the flying frisbees or bonfire parties. Though these activities are too far away to bother us here, they do raise issues. Western writer Roderick Nash, author of *Wilderness and the American Mind*, points to the tendency in accessible backcountry areas for people who just want to recreate outdoors to drive out people who seek a wilderness experience. "The change parallels the evolving nature of river running everywhere from a solitary, risky, expeditionary activity to a form of mass recreation." All of us could ask ourselves, "Is this something I could do in a city park?"

The banks of the San Juan are not, like those of the Green or the Dolores, lined with tamarisk, willows, or scrub oak. There's no thin green line, yet the river corridor does support small trees. Netleaf hackberries with characteristic galls on their leaves droop over a drainage. Under the lip, its trunk twisted by deluges, a bright green box elder also hangs on. In the Southwest, where the sugar maple does not grow, both Indians and white settlers tapped box elders for their sugary syrup.

Just six feet away, his pale yellow head cocked to look me over, a colorful collared lizard does pushups. Such movements, it's surmised, either keep these lizards cool or signal that "this is my territory." This flashy fellow is the bird of paradise among reptiles. Behind his jet-black collar, he becomes an almost phosphorescent green with yellow speckles on his back. His forelegs sport yellow gloves with black stripes on their backs, while his rear legs shade toward turquoise.

The smell of gasoline reminds me that oil oozes from these banks. When he became the first to run the San Juan from Durango, Colorado, to its mouth in 1882, E. L. Goodridge discovered these oil seeps and filed the first claim for oil. Decades later geologist Herbert E. Gregory noted, "A sand bar is used by flies as a breeding place, and hundreds of larvae live in the oily substance." This place is getting to me.

Though I wish it were otherwise, I catch myself seeing my mood reflected in the smudged stars. It's some consolation to recall Robert Frost's "Desert Places," a meditation on the voids between the heavenly bodies: "They cannot scare me with their empty spaces/Between stars—on stars where no human race is./I have it so much nearer home/To scare myself with my own desert places." The silence of infinite space reflects the void within, the silence I hear tonight.

In the morning we enter a cove at Slickhorn Gulch, named after a breed of longhorns that once grazed this canyon. On days when it's not overrun by river runners, lower Slickhorn is one of the gems of the Southwest. But on hot summer afternoons its plunge pools can begin to resemble a spa: bank-to-bank bodies, thumping headphones, and suntan oil floating on the roiled surface. There are six unnamed pools in lower Slickhorn Gulch, and this is surpising. We humans like to create and manipulate symbols—an ability we hold to be unique among our species—to affirm our humanity and dominion. We accentuate features we find compelling and, by doing so, create a sense of belonging in a place. The Pueblo Indian cultures of New Mexico weave a complex association by naming individual rocks, trees, or other features of their landscape, so that even someone who hasn't lived on the land senses the bond—a feeling of relatedness.

But naming has its risks. Too often we invent inappropriate terms such as "killer whale" for orcas or, for such a majestic mountain range as the Grand Tetons, "the Big Breasts," substituting human interest for the ability to see things squarely on their own terms. Personification can be comforting, but when we project ourselves too much onto nature, we foster the idea that the world was made for our pleasure. Naming also feeds the impulse to pin down nature like a bug on a board, giving the illusion of understanding where there is none. But today compromise seems best to me: I don't feel compelled to name every pool in sight, but I won't resist when places whisper their names.

Remarkably, to my delight, Slickhorn is deserted. Our hike begins on a historical note as we pass the rusting wreckage of an oil drilling rig left from the 1950s. To the right loom the switchbacks that descend from the canyon rim, far above the river. Years after the gold rush he helped to create, and immediately after his first well turned out to be a gusher, E. L. Goodridge constructed this road in 1909 to bring in oil drilling equipment. Using pack animals, he hauled machinery all the way from Gallup, New Mexico, and eased it down this track only to see his equipment break loose and smash to smithereens. His loss, nature's gain. Given the dark deeds that lust for black gold has encouraged in the twentieth century, it's tempting to dance on its grave. Then it strikes me that we've just tied up a rubber boat with nylon rope—and that rubber and rope no longer come from plants.

As the air shimmers, Slickhorn's pools serve delicious sensory treats. Yellow columbines and maidenhair ferns work their magic on Barbara. She toes the water and dives in, yelping half in pleasure, half in pain. I recline on the damp, cool sandstone near a drip to savor its spray. Deeper under the overhang, a red-spotted toad sits on a mossy knoll, throat pulsing. The toad is mostly buff but with an important addition: waxy orange spots behind the eyes that offer protection. When a predator chomps down, these spots squirt secretions that inflame its mouth, throat, and stomach.

These pools teem with life. At the edge a brown-speckled spider waits for the water striders that, so far, stay out of danger as they skate over a hair. Deeper down a dragonfly nymph first scratches its head with its feelers, then engulfs a large hunk of algae. Moments later a puff of sand reveals the formidable claws of this mud-colored dragon as it grabs a tadpole. The struggle is short. Another polliwog wriggles its way toward the surface, twisting its head as it gulps in air.

Two kinds of shrimp inhabit the pools, one true, one misnamed. Freshwater shrimp rest on a twig, their tiny gills fanning the water and stirring up the ooze. In contrast to these fairy shrimp, which are translucent and swim gracefully on their backs, tadpole shrimp (or notostracans) look like miniature horseshoe crabs. Notostracans are actually not shrimp nor, aside from having a bulbous head, do they share much in common with tadpoles. More closely related to extinct trilobites than to most saltwater shrimp, these primitive crustaceans are living connections with the earth's prehistory. One species, in fact, has survived unchanged for two hundred million years. It's astounding that a species adapted so perfectly to its niche that subsequent mutations apparently offered few advantages.

Nearer to the surface the backswimmer beetles jerk and stop. Their oar-like hind legs, flattened and fringed with hairs, shoot them backward. Others hang upside down at the surface, where they listen for prey and replenish their oxygen supply. Since they carry air inside their bodies, they can stay submerged for hours. However, this stored air also makes them float to the surface, tummies up, when they don't propel themselves down regularly.

Skimming just above the pool, much like planes in holding patterns, two cinnamon-orange dragonflies patrol in straight bursts. Their acrobatics, however, far exceed anything possible in aviation. Dragonflies can catapult themselves straight skyward and carry fifteen times their own weight. When they encounter a swarm of gnats, they easily outmaneuver the most agile of birds. Though adult dragonflies possess extraordinary vision and formidable jaws, one rarely sees them seize their prey. They grab a gnat or mosquito in one of their agile darts but spend the majority of their time cruising. These two dragonflies fly in tandem, buzzing each other playfully every few moments.

While the sun broils the rocks nearby, we huddle under a

dripping ledge just far enough from the spray to build up goose flesh. As soon as we're chilled to the bone, we make our move for the next oasis. Once on the path, our pre-cooled skin enables us to handle the sun. Desert heat and light animate us as we skip along the way. This is pure sensation, total kinesthesia enhanced with the ego buzz of beating the game by staying cool amid all this heat. Pants still soaked, we head up and around the plunge pool. From far above we can hear every drip. Below lies Echo Pool, as I call it. Under its overhang, just downwind from the spray, maidenhair ferns trim the seeps in the sandstone, their fan-shaped fronds offering lushness in the desert.

Slickhorn also hosts other plants that aren't common along the river: lavender penstemons, orange paintbrush with lime-green pistils, healthy buffaloberry bushes with water-bloated leaves, and sacred datura, the most showy of the Southwest's night bloomers. Pollinated by night-flying insects such as sphinx moths, these drooping white trumpets are the Easter lilies of the desert. Several Indian tribes have used this remarkable plant to relieve pain and cure wounds. Given its concentration of narcotic alkaloids, datura often energized ceremonies intended to induce visions. Like peyote cactus, though, these beauties are both hallucinogenic and poisonous. Even the datura beetle, the sole insect to feed on this plant, can be poisoned by its leaves and stems. Paiute shamans prized these plants but understood that they could both "make one sleep and see ghosts."

On our way down the limestone ledges of Slickhorn, we take one last private dip. Greenish bands of light radiate from Barbara as she strokes the water, breasts bobbing. When she tries to evade me, we cavort like dolphins, nipping each other like amorous otters, liquid and weightless as the universe enters us. Minutes after we've whirled the pool, while its warm ripples are still lapping, we feel a resonance to our love making. Naming this pool would be like learning lyrics to a Beethoven symphony and then wondering why you couldn't hear the harmonies.

Later we encounter the gang from Outward Bound. Water ouzels bolt from their nest beneath the cliff, unnoticed by these athletes. While one guy clings to the precipice, using the pool below as a safety net, others cheer him on: "All right, Dave!" "O.K!" "Way to go, Dave." This cadre of twelve will hike upcanyon, scale a wall, repel a cliff, and then march down Grand Gulch.

Back at our campsite chukars—the colorful quail from Asia—are putting on a real show. Their loud clucking comes from the cliff behind us, but we see nothing but rocks. Finally something raises its head and paces on the ledge between clucks, strutting into full view. The clucks louden into feline squawks. Two chukars, black stripes on their wings and black necklaces showing boldly, call from boulder to boulder. Then, wings whirring, they streak right above our heads. Like ring-necked pheasants, chukars are introduced game birds. So far, the effects on native ground birds seem minimal.

Head down, dejected, Dave from the Outward Bound group trudges toward the beached sportyaks. Berating himself as a quitter, he's decided to leave the group. "What's wrong with quitting when you're ahead?" Barbara asks. Dave has conquered his fears of heights, so he doesn't need to keep proving himself. From our point of view he's shown the courage to break away from the group pressures to use nature as an obstacle course. This may take more guts than performing physical feats.

My first attempt to start a fire fails. It's been too long since

I've built one, so I have to relearn the art of using small kindling and leaving plenty of space for air. Finally the scent of cedar spices our campsite. Our small fire forges a link with the thousands of generations before us who also sat around their campfires, protected from the dark. It connects us with the Anasazi who burned juniper to the point that they eventually denuded hillsides, hastening the erosion that probably contributed to their decline. Bony flames like writhing arms plead with us to learn from the ancient ones.

As the wood crackles, our world contracts. Footprints pock the sand, then step into the unknown. Larger-than-life shadows close in as the flames die down. In the evening's last moments our world becomes throbbing red embers and pin-prick stars. Stretched out on the sand, we become children still a little afraid of the dark. Yet to snuggle up to the coals, avoiding the vastness and darkness beyond, is to miss the starry heavens that so moved the young John Keats: "I must have a thousand of those beautiful particles to fill up my heart, . . . I do not live in this world alone but in a thousand worlds. . . . I melt into the air with a voluptuousness so delicate that I am content to be alone." Keats, like many nature lovers, may not have found companions who were compatible with his contemplations.

A flashlight bobs in the dark. Barbara pants that a "something" is rustling like dry leaves. Rattlesnakes hunt at dusk while the warm sand keeps them active as mice and toads come out. Barbara plays her flashlight around once more, then checks the zipper on the tent. Scoring a point in a longstanding controversy, she touts the advantages of sleeping inside. Granted, there's no worry about snakes in the tent, but it's a shame to miss the moon, the stars, and the wildlife. I remind her of the baby spotted skunks, soft as kittens, that once romped around our sleeping bags in Grand Canyon.

"Right, and you jumped when one ran up your leg." She has the last word for the night.

After breakfast we push off for Grand Gulch. Silent and shallow, the San Juan slows and widens. Not a single bird sings. Maybe it's Sunday and the birds are observing the Sabbath—anyway, it's freeing not to know the day of the week. Squeaky oarlocks defile the quiet until I muffle them with lip balm. The cliffs become increasingly red, darkened with streaks of desert varnish. We glide into a small cove, tie up, and scramble up the rock benches. Two ravens croak as they glide into their nest.

Here at its mouth, fifty twisting miles from its origin, Grand Gulch looks different. Trapped behind giant mono-liths, driftwood has been piled high by flash floods. The great cottonwoods of upper Grand Gulch don't make it down here, at least not in one piece. Pulverized by raging water, their twigs end up as nestlike wreathes stuck in the brush. Early river runner Bert Loper recalled an impassible rapid at the mouth of Grand Gulch, but since Loper's time the river has rolled the immense boulders out of its way.

As we hike on the broad, sunburst canyon floor, we guzzle water to counter the heat. Grand Gulch is rough and raw, no place to run low on either bodily fluids or trace elements. Sweat soon stings our eyes, and even our sunglasses feel too hot to wear. While we rest in the shade, Barbara tells me not to breathe on her. We've had enough of the heat so we turn back toward the pool, just above the river.

As we sit there silently, a great blue heron makes one whuff of its great wings and splashes down. We freeze, barely breathing. The heron wades patiently, head extended, dip-ping its beak and shaking its rubbery neck. After minutes of staring into the pool, it lunges two steps and one wing flap forward to spear a small sucker. Then it takes three giant steps back to the sandbank, where it gulps down its catch. With no split between hunter and hunted, the fisher seems all fish.

As a sand-colored toad hops, the heron strides toward it and cocks its head. But then it also spots us and freezes. Suddenly the heron extends its neck, spreads its slate-colored wings, beats the air, and squawks in protest. Its dull yellow legs trail and bob behind like folded landing gear as its neck snakes into an S. This great bird goes "grak, grak, grak" as it tries to cross the wind but gets swept upriver, feathers ruffled, legs blown askew.

The wind reminds me that the takeout is thirteen river miles away—and that upriver gusts will probably require constant rowing. It's usually a tough, slow go. We float by Oljeto Wash where John and Louisa Wetherill established an early trading post in Navajo land. Today two Navajos in long sleeves and dark hats walk down to the river; one carries a fishing rod, another a tackle box. This intrigues me because traditional Navajo (or Diné) usually observe their tribal taboo against eating fish. To an outsider, this prohibition might seem odd for a people who emigrated from the Great Slave Lake, a great fishery. In their six to seven centuries in the Southwest, however, the Diné have evolved a new cosmology literally embedded in the landscape. As Robert McPherson notes, many traditional Diné believe that fish come from down deep and thus, like the earth, should not be disturbed.

Miles of flatwater test our tenacity. The fact that Lake Powell has silted up while the river flow has decreased definitely exacerbates a big problem: sandbars. Since the lake has dropped several feet since the mid-1980s, the San Juan is now cutting through the silt it deposited years earlier.

To make matters worse, this stretch is also a notorious wind tunnel. The river runs almost exactly west and funnels winds that sweep unimpeded across Lake Powell. To complicate matters further, the river is sometimes too shallow for oars—the raft becomes easier to tow than row. In recent years more than one rafting party has arrived hours late, blasted by gales, after slogging for miles through knee-deep water. I wonder if we'll make it before dark.

To take advantage of our light boat, I attend closely to the current. Sometimes it runs along one bank for a long stretch only to sneak across a sandbar to the other side. This leaves us in a backwater, buffeted by hellacious winds. Reading the water gets tricky, for low water, sandbars, and winds all wrinkle its surface. Moreover, the normal flow signals have changed. In deeper water, or under still conditions, rounded boils typically indicate a slowing current; here they suggest that some current is still flowing. Similarly, riffles typically indicate an accelerating current; here they signal slow, wind-blown water.

As if all these wind-and-water dynamics were not complex enough, fatigue and discouragement mount when we struggle to row past the same rock twice. Desert air parches my mouth. When Barbara stands to call out the sandbars, she adds to the wind resistance. My hat flaps in my face, brim flat against my sunglasses. Unable to see much anyway, I concentrate on stroking and breathing, heaving and ho-ing. Faster strokes reduce the wind's ability to slow the boat but soon, given the frailties of the flesh, they degenerate into slaps at the surface. When the river deepens, I submerge the oars like underwater sails. If there's no current, at least they help to keep us from getting blown upstream.

Barbara's visor blows into the river. As we reach for it, the wind shoves us onto a sandbar. We'll have to haul over this one. Barbara pushes while I tow our barge like Bogart pulling *The African Queen*. Despite all this travail, the riverbanks are alluring. The canyon walls become more rounded, varnished, and colorful. I look over my shoulder as I row, hugging the bank while watching for snags from the corner of my eye.

In the distance, so near and yet so far, the gray shale of Clay Hills takeout beckons like the Cliffs of Dover. To take my mind off the interminable rowing, I watch the lush banks which teem with Russian olives, willows, beeplant, and even bracken ferns. Beavers don't attempt dams here, of course, but other signs of them appear everywhere: holes a foot or more in diameter, prints and claw marks in the mud, chutes where they splash into the water. In a backwater a flock of yellow-headed blackbirds clings to the swaying cattails. Even in gusts like these the males fly with their bodies cocked upward to display their flashy heads.

Although there's a channel again, this last mile is shallow and tricky. At best, the oars jam on the bottom, slowing our movement. At worst, they pop from their locks, saved only by their safety cords. In either case, I spit the appropriate nautical epithets. I ache and don't care how we get to the end.

Finally we wobble ashore. Everything—frame, raft, cooler, oars, ropes, baggy pants—is encrusted with mud. Our legs look like that heron's; my hands resemble that lizard's paws. And my case of "boater's butt," a combination of heat rash and sand abrasion, is now acute.

While Barbara crams everything into the van, I strap the frame to the top. We face seventy-five miles back to Mexican Hat for sustenance. My feet lounge on the dash as my scruffy head leans back. Wind and water have pushed us to our limits and soothed our minds. Muddied, sanded, parched, and windburned, we don't need a watch to tell us that we're famished and exhausted. Played out, yes, but also rejuvenated.

Interlude

Dammed Nearly to Extinction

Though their numbers are greatly reduced, native fish still swim in the Colorado River and its tributaries. But because of human disturbances, their existence hangs in the balance.

The Colorado squawfish, the largest member of the minnow family, swims great distances from whitewater canyons to backwater nursery areas. Today, however, squawfish seldom reproduce because dams intercept the cold spring runoff that triggers their spawning. Once common enough to catch commercially for fertilizer, the squawfish now teeters above the abyss of annihilation.

Another native fish, the razorback sucker, once grew to lengths exceeding three feet. One of the largest suckers native to North America, it now appears on the threatened species list. A few razorbacks still ply the Green River, where they've faced the fluctuations caused by the Flaming Gorge Reservoir, and razorbacks also inhabit the Yampa River where a free-flowing stream provides ideal habitat.

Indigenous chubs are also in trouble. The humpback chub, another member of the minnow family, seldom weighs more than two pounds or exceeds a foot and a half in length. Though formerly common in the lower Colorado, this rare species now lives mainly in the Little Colorado River, another free-flowing tributary. The bonytail chub, the rarest of these fish, has declined to only a few individuals with no known reproducing populations in the wild. Its close relatives, the Virgin River chub and the woundfin minnow, inhabit only the Virgin River, one of the most jeopardized streams in the Southwest.

Throughout the Colorado River basin the main causes of such die-offs are dams and diversions. Native fish require specific natural conditions to spawn, and dams not only reduce downstream flows but also, as they release cold water from their reservoirs, eliminate the warm water that some species require. In addition, the dams lead to increased pressure for sport fishing in reservoirs. Many native fish cannot compete with the preferred exotic species.

What difference, some might ask, would it make if these fish went extinct? It's usually difficult to muster public support, let alone reverence, for animals that are not symbols, not game species, not edible, not cute, or not seen. Yet these native species belong in the rivers where they've evolved, and they need our help to survive.

Another reason for preservation is self-interest. In 1980 many might have asked why we should protect the Western yew, then an obscure tree with little known utility. Today we understand that this remarkable plant is the source of taxol, one of the few treatments for ovarian cancer. Early proponent of environmental ethics Aldo Leopold remarked that "To keep every cog and wheel is the first prerequisite of good tinkering." We need to approach the future with a gene pool that's as large and varied as it can possibly be.

As biologists document the damage in Grand Canyon and elsewhere, perhaps the public will no longer condone costly dams. The West has long endured an almost imperial system of interlocking elites including the Bureau of Reclamation, the Army Corps of Engineers, water conservancy boards, power companies, and big business. In one of the great ironies of today's West, the same people who condemn federal management of public lands often turn around and promote federal subsidies for their water projects. But as Westerners have become more aware of how this subsidy system works, things have begun to change. With fewer federal funds available for new projects, dams are becoming much less appealing to local planners.

How can Western rivers both host native species and also offer resources for people as well? The answers are complex, but they lie with more careful operation of dams, more resourceful water conservation, recreation compatible with natural values, and populations appropriate to the carrying capacity of the land.

13.

White Water, Black Rock

The Colorado is an outlaw. It belongs only to the ancient, eternal earth. As no other, it is savage and unpredictable of mood, peculiarly American in character. It has for its background the haunting sweep of illimitable horizons, the immensities of unbroken wilderness.

—Frank Waters, *The Colorado*

As the sky clears, snowmelt from the Rockies swells the Rio Colorado, "the River Red." Sun and sky, earth and river come together on a vibrant June day. Beneath the old bridge, barn swallows skim the river's speckled surface. Scissor tails fanned as they hover, these lovely birds daub their faces into the mud, then fly straight for the bridge. There they shake their heads to ease the daubs onto a cup-like nest. Their satiny black bodies and rusty throats gleaming, males and females take turns molding the mud with their bodies, using their beaks to smooth the cusp.

This will be a four-day, eighty-five-mile solo from Fruita, Colorado, to Sandy Beach near Moab, Utah. It was from this very spot, fully a decade ago, that my friend Jeff and I launched our first float trip. But this time the *Grey Coyote* will row it alone. The main challenges will be the landings, which are tough for one paddler, plus the Westwater Canyon stretch where I'll meet friends to run the rapids. Since the Colorado River skirts one of the largest wilderness areas in the Southwest, from here on I'll rarely see even a rudimentary road.

Once the river leaves the Interstate, the riverscape comes to life. A gust silvers the Russian olives, then carries the sweetness from their pale yellow flowers. Introduced as a windbreak, Russian olives have spread along the river corridor. A spotted sandpiper bobs among the reeds, uttering high-pitched cries as she feigns a broken wing to draw predators away from her nest. Soon she takes flight, wings beating stiff, shallow strokes just above the water. After alighting, this sandpiper dashes along a sandbar, legs a blur.

Black-billed magpies scold and sail between the fragrant trees, buoyed by long, wedge-shaped tails. Large white patches flash from their wings, while their iridescent tails stream behind with hints of emerald, indigo, and burgundy. Though spectacular, magpies are a much-maligned bird in the popular imagination. As recently as the 1930s Westerners attempted to exterminate these birds, often because they were falsely accused of destroying crops and even of pecking out the eyes of calves. When ranchers saw a magpie on a cow, they seldom understood that it was probably eating flies or maggots from wounds left by the branding iron. Thus magpies fell victim to ignorance—and to the human tendency to find scapegoats.

These utilitarian views contrast with those of Southwest Indians. One Pueblo story recounts how two sisters once competed for the first rays of the rising sun. The older one asked a magpie to head east and eclipse the younger sister from the sunshine. But the magpie stopped for breakfast at a carcass, which soiled its plumage. When this happened, the older sister became incensed and decreed that magpies would be dappled and eat dead meat. While this story illustrates how Indians often saw animals as emblems that revealed the ways of the natural world, it also suggests that they, like other peoples, sometimes cursed species that defied humans.

On the far bank, behind the olives and cottonwoods, rise the red-rock bluffs of Colorado National Monument. Far above, set against billowing white cumuli, rocky peach-hued fins glow in the afternoon light. These smooth outcrops provide a glimpse of Rattlesnake Canyon, a magical place that the BLM tries to protect from over-visitation.

The slickrock outcroppings of Entrada sandstone tend to round, cup, and bowl. Though gritty winds abrade these surfaces, most of this sensational sculpturing results from "solution," whereby water containing carbonic acid dissolves the minerals that cement the sand grains, or from "chemical sapping," whereby dissolved salts crystallize on the surface and pry away grains of sand. Physics and chemistry aside, the smooth shapes and pastel color of Entrada sandstone place it among the loveliest rocks anywhere.

A huge chunk of bank crashes into the river, raising a geyser of spray. A beaver stops to stare, motionless, as the *Canyon Wren* glides by. As two other beavers swim ahead, one slaps the river hard and dives. The other, hearing the distress call, arches its rump but doesn't slap.

My raft drifts up to a fellow standing knee deep where Salt Wash enters the Colorado. Dressed in baggy field clothes, Pablo bends to net aquatic insects where the turbid waters of the creek meet the silty flow of the river. A grad student in zoology, he explains that sand and gravel not only help the river cut its channel but also weigh down caddis fly grubs so they don't get swept away. Peak flows, he points out, stir the gravel and keep the bottom accessible to insect larvae and small invertebrates. Without this regular flushing, sediments build and harden, rendering the river bed impenetrable, or they may clog backwaters where fish need to spawn.

Just below Salt Wash, Horsethief Canyon ends and Ruby Canyon begins distinctly. As I pull against the stiff breeze, my eyes glimpse the tunnel where, years ago, a fast freight almost totaled Jeff and me. Here the grayish, greenish, and purplish Morrison formation, the West's great boneyard for dinosaur fossils, yields abruptly to the uplifted red sandstones of Ruby Canyon. Curved and bent layers suggest faulting on both sides of the canyon. The Colorado follows the fault line, exposing different rocks on the north and south canyon walls.

On the westward stretches, riffles sparkle where gusts whip them into spray. As my arms fatigue, I land where dunes afford some protection from the now relentless wind. The sand is far too hot for bare feet. Bowed over by the breeze, waxy-yellow bee flowers trace sandy crescents on the dunes.

Dusk falls slowly. Cottonwood leaves patter like rain; then, hanging limp like apples, these giant hearts click no more. Doves coo and gnat-like midges whine. The slightest puff blows a cloud of gauzy-winged ephemera backward, only to let them drift forward when it wanes. A moth buzzes by. Waxy leaves now glint in the moonlight, and, infinitely far beyond, the sky sparkles like a moonlit beach. Quartz grains twinkle as my toes wriggle into the warm sand. My whole being participates in the universe, near and far.

As I arrange my nest for the night, a tiny whisk broom reminds me of Barbara, who used to sweep out our tent on our river trips together. As sand runs through my fingers, I wonder if true communion with people is like the grains that evade my grasp. Now, bereft of dependents, I can take more chances with life and run a river alone.

As another desert day breaks, the sky turns baby blue. Since the river has risen during the night, the silver, blue, and yellow *Canyon Wren* basks in an eddy, awash in the brilliant sunshine. Sun and wind have begun their diamond dance on the river, and I do a yoga sun salute from the bank. I resonate with a world that plucks the heartstrings of my being.

After a snowy egret splashes down, one of its yellow feet stirs the shallows for prey. Minnows fret the surface of the backwater. Since most of us associate egrets with subtropical wetlands, it's surprising to see them here in a near-desert. But egrets have expanded their range and small populations frequent rivers throughout the Southwest. From a clump of willows this egret rises slowly just three feet above the water, a snowy shape set against the deeply shaded canyon walls. Once decimated to embellish women's hats, snowy egrets have enjoyed federal protection since 1918. As a result, they've made a remarkable comeback.

Aware that the wind will rise soon enough, I drift past Mee Canyon, a wonderful hike offering striped spires along the way to a giant overhang carved by a spring. Black Rocks, a memorable landmark on this stretch, appears just downstream. Along a distinct fault line the ancient precambrian schist and gneiss surface here, buffed and fluted well above today's waterline. By narrowing the river and increasing its speed, these hard metamorphic rocks are subjected to accelerated carving and polishing. Currents boil up from an irregular bottom. At high water these hydraulics can become treacherous, for whirlpools can suck down careless boaters much as they do in Westwater's rapids.

Across the river the California Zephyr streaks by, a silver flash along the rusty mudstone and shale of the Chinle formation. Seated on a mossy rock, nude, I wave to the train's passengers. Not surprisingly, the best response comes from the club car. Many years ago, riding this train home from the Summer of Love, I first beheld the red rock country in this very canyon.

The sun throbs hot and dry in Moore Canyon, a great hike from the river. Along its drainage the single-leafed ash trees dangle their light-green fruit. Wings quivering, a black-and-yellow tiger swallowtail butterfly lands on pink milkweed. With blue dots, yellow eyespots, and orange and black bands, the green caterpillars of this large butterfly are also striking. In part because their larvae feed on willow, alder, and poplar trees that are widely distributed, these large butterflies are more abundant than, say, the sleeker zebra swallowtail, whose caterpillars eat only the foliage of the pawpaw tree.

The sun scorches the smooth bedrock beneath my sandals; only the slight breeze makes it possible to hike under the noonday sun. I find shade on a bench of rust-red Chinle shale. Amazed by the reds, purples, browns, greens, grays, pinks, whites, and even blues, the Navajos called the Chinle "the land of the sleeping rainbow." Where water percolating through the porous sandstone reaches this shale, it sometimes hollows out pockets suited to the human body—sinuous, grainy places to rub the earth. From one of these I observe the wind that streams over the bench, shortcutting to the inside of this twist in the canyon. Air and water flow quite differently: one seeks lower pressures along the shortest route, the other, so much heavier, pushes toward lower terrain where, due to its momentum, it rides the outside of a curve. Yet both rise and fall, crest and trough, with the rhythms of the earth.

Krok, krok. Two ravens spiral from hundreds of feet above, feathering the updrafts. As one approaches a nest, squawking and croaking echo from the clifftop. Streaks of white guano mark this nest. Ravens repair their nests year after year, and when they finally abandon them, hawks and owls often move in. Similar stripes of whitewash mark another nest on the opposite wall, for ravens build more than one nest in their territory so they can alternate between roosting sites. Smart birds.

Like magpies, ravens have long endured discrimination. Hebrew legend reveals that when Noah released a dove and a raven, the dove came back with an olive branch, a sign of peace, but the raven flew the coop. Doves are really pigeons, birds that flock toward human habitats, but ravens keep their distance. Since they won't nest around people, perhaps their worst sin is that of disdain. We humans don't accept snubs from the fowls of the air.

Since ravens are as intelligent as a smart dog, they may also challenge human delusions of superiority. Rather than running mainly on instinct, ravens think through new situations. They drop clams to break their shells and even invent games in which an airborne bird will release a stick while others try to grab it before it hits the ground. Many zoologists credit ravens with insight, the ability to image a solution rather than to simply learn from experience—and the sort of intelligence we've always reserved for ourselves. But these ravens also embody the primal. Strange tapping sounds, guttural croaks, and black plumage blend with white stains, fallen sticks, red rocks, and blue sky. Dark omens against a blazing sun.

On the river bank other boaters lounge under a canopy. Three other boats will also face Westwater Canyon's rapids. That's reassuring.

Violet-green swallows with satin backs skim the river, their beaks leaving tiny wakes, then shoot straight up into the wind. Since these birds do not typically nest in colonies, they must defend the area around their nest holes from other cavity-nesting species. However, sometimes violet-greens help Western bluebirds rear their nestlings and then, once the bluebirds have fledged, breed in the same nest. Nature is infinitely more complex than the view of incessant competition among species.

Where the river bends right, the Utah state line is indicated by "Utaline" lettered behind the railroad. Out here, far from civilization, such boundaries seem absurdly illusory. A gauging station here tells state engineers how much water has been wasted or delivered, depending on one's provincial viewpoint. Many Coloradans, for instance, think that any water that has flowed out of their state has been foolishly squandered.

Utaline sits just four miles upriver from Westwater Ranger Station where tomorrow I'll make my connection with friends. This is a good place to practice a landing, one where I pull in close, drop my oars, spin, spring over mounded baggage, grab the rope, and jump for the bank—all in one motion. But I fumble the rope and run into a tamarisk bush. Not the confidence builder I needed.

While I slouch on the wet sand, swigging from a wet canteen, a great, thick-winged bird soars far above the canyon wall. Mottled with white on its underwings and tail, its wingspan nearly seven feet, this is definitely a young bald eagle. These astounding birds are not thriving along the Colorado, however, despite some special assistance. When the prairie dog population in the nearby Cisco Desert crashed, Utah wildlife officials, BLM rangers, and river outfitters set out bald eagles' favorite food: fish. This was a stopgap measure, for the real problems are the loss of large cottonwoods for nesting and the disappearance of live fish and rodents. Diminished river flows, water pollution, fewer trees, the extermination of prairie dogs, and heavy fishing all cause eagles difficulties.

Nationally, though, the bald eagle has reversed its decline toward extinction. Ornithologists credit the bald eagle's comeback mainly to the banning of DDT which causes raptors to lay eggs with shells so thin they break before they hatch. Other causes of death, including electrocutions on power lines, accidental poisonings by predator-control agents, shootings by gun nuts or by ranchers, and poaching by traffickers in body parts have abated, though they still take their toll.

As the sun enlivens the cliffs across the way, each swirl, each ripple made by bird or bug or fish shows up as a sky-blue ring on the river's orange surface. A great blue heron labors along the far shore, the downbeats of its great wings dipping into the fluid light. When another raft plows the surface, the river makes room, accepting the dimples left by the oars, and then erases the wake as if the boat had never been. The man and woman wave a knowing hello and goodbye. They're trolling, attentive to signs from the river's brown depths.

When I sink into anxiety about the rapids I face in the morning, nature provides perspective. Well camouflaged in the dappled light, a tiger beetle sprints on hot sand to seize a cricket. After its jagged jaws puncture the carapace of its prey, its other mouth parts suck the cricket's juices. As I lean in for a look, this tiger beetle's defense proves as impressive as its offense. Once its huge, bulging eyes detect my slightest movements, it takes flight, zigzagging over the water. But the ferocious predator itself becomes prey. A black phoebe darts from a rock to intercept it in midair, then shoots back like a crow carrying off a mouse. The cricket and the tiger beetle both existed just moments ago, no less alive than I am now.

Are insects really mindless robots? Can their complex behavior be reduced to mere instinct? True, their brains are much smaller than ours, but does that mean we can deny them all the consciousness that we reserve for ourselves? Famed paleontologist Stephen Jay Gould asks an even more troubling question: Is human consciousness a mere accident? Let evolution play out again, he contends, and "the chance becomes vanishingly small that anything like human intelligence would grace the replay."

However limited the consciousness of animals may be, at least it's not disconnected from the world. As we evolved language, we may have lost touch with other species as we began to talk only among our own kind. When literate civilization developed with the speed of the written word, the silent gap between us and other living beings probably widened. And when we evolved the heightened awareness that led to a sense of self, we apparently did so at the cost of participation with nature. For us, the paradox is that coming to consciousness and developing language may have ruptured a previous wholeness with nature and divided the Self from the Other, from the outside environment. Estranged from nature, we live in boxes, fearful of what we can't control.

Tomorrow, for instance, the river takes charge, and that scares me.

As I wash my pans, Westwater's rapids roar in my mind's ear. Last winter I yearned to be challenged, even overwhelmed, by nature. But now, as I face the actual challenge, I'm less interested in relinquishing control. River runners say Westwater is nothing to worry about, yet with names like Funnel Falls, Skull Rapid, and the Room of Doom, I do wonder. Rock climbing, car racing, and sky diving have never attracted me. I'm not addicted to adrenaline rushes, but I do crave the opportunity to immerse myself in the river's Great Flow. If you want this full relationship with a river, then you need to float long stretches, which often means that you have to face rapids.

As usual, the river offers good nightlife. At dusk nighthawks join the early bats. Though both may hunt the same prey at the same time, few conflicts occur. Nighthawks freewheel, changing gears to shoot forward with quicker, erratic wing strokes. These aviary acrobats usually fly silently when making their steepest banks and turns, though they can make unsettling sounds. Most typical are their high-pitched cries made to each other in flight. Another is a high-speed snap apparently made with their beaks. Still another, more startling sound is a loud *graaate* plus a *whuff* made when one bold bird jets just above my head, and then, dragging a feather, zooms upward with a sudden deep *whirrr*. As the "booming" nighthawks quiet down, tiny pipistrelle bats flutter erratically.

It's not the night boomers, however, that make sleep difficult; it's the rumbles, squeaks, and hisses of trains. This is one of those fitful nights when you wake, wondering if you're getting any rest, only to realize that you must have slept because you've had a bad dream. As a result, I oversleep.

Fretful that I'll miss my friends, I gulp down instant coffee as I pack up. Double-bagged binocs and camera disappear into dry sacks. Netting battens down everything.

I'm afloat by eight. My rendezvous point, the Westwater put in, lies an hour downstream. I row hard, relishing the bounce from oars made of New Zealand pine. Muscles warm as rhythmic breaths loosen my tightened chest. Behind the shiny railroad tracks, splendid sandstone panels slide by, unappreciated. It's not the intrusion of industrialism that separates me from the rocks; it's the hurry-worry monkey chattering in my ear. Right now I'm too goal-oriented and worried to live in the present.

A hearty boatwoman hails me from the dock. Betty, a seasoned raft guide, will captain the lead boat. Chris and Will, mellow medics and bon vivants, arrive back from running their shuttle with Leif, a passionate paddler who teaches English. Betty and Leif describe the Westwater rapids as "sporty." The rest of us wonder what that means.

Speckled by an armada of driftwood, the Colorado runs high and tan. Its currents finger new channels through the tamarisk, course through willows, find old channels, and pick up the pulverized twigs and bark it deposited in years past. Scum spirals like nebulae in an eddy. While spring runoffs no longer create enough backwaters for many native fish to spawn, they do bring food to fish, nutrients to plants, and, we hope, carry boats over boulders.

After fidgeting enough to prepare an expedition down the Amazon, the gang gets ready to go. The first five miles are quiet, almost too much so. On the left is Wild Horse Cabin, once home for outlaws, rivermen, and prospectors whose broken sluice boxes languish just downstream. Widening out as the canyon opens, the Colorado ponds behind the Uncom-

pahgre Uplift, through which it has incised Westwater Canyon. At Little Dolores beach, Leif picks a blue thistle for good
luck before we push off for Outlaw Cave. There, under a
blackened ceiling, sit a potbelly stove and two rusty beds.
Antique bottles litter the floor. Legend holds that two brothers robbed a bank in Vernal, Utah, and later hid out here for
eighteen months around 1913. Back then only a few intrepid
souls would have ventured this far, for the rapids had already
splintered several boats.

In fact, Westwater was long known as Hades Canyon.
Cut into precambrian rocks nearly two billion years old,
Westwater is a three-mile gash in a massive bench of sandstone. As the molten sun glares off granite-veined, fluted
black schist, it resembles the Inner Gorge of the Grand
Canyon. The biggest rapids occur where the river exposes still
harder lava intrusions into this already resistant rock. Where it
dives into this defile, the river narrows from three hundred to
less than one hundred feet across.

The time has come. As jagged walls close in, we cinch
our life preservers, further constricting our already tight
chests. With Betty in the lead, we enter the narrowing gorge.
Marble Canyon Rapid gives us a good ride with just enough
spray to speckle our sunglasses. Chris and Leif click paddles in
the air, blithely assuming that my oars alone can power us out
of the current. But the flow is surprisingly swift and it requires
all my strength to propel us across the sheer line, where we
finally eddy out. As soon as we enter the eddy, Betty immediately pulls back into the current, on her way to Funnel Falls.
There's no time to scout or scheme: we'll have to follow our
leader.

The channel enters the dark gorge. Where an angry
volcanic intrusion scars the sheer cliff, the river enters the

Underworld: the Rio Colorado becomes the River Styx. Accustomed to rapids that are spaced, I assume that there'll be time between them to bail. But Betty's boat barrels right into standing curler waves that toss boaters like Raggedy Ann dolls. The skewed "V" of the tongue licks hungrily into the wild water on the right, so I stay mid-stream. The guttural roar reverberates between the walls as we all strain to read the water. The river seems just to end—only froth is visible beyond the drop.

Shit. Funnel Falls is really a spillover dam. Worse, the accelerating current now makes it impossible to reach the tongue. The smooth flow bulges up in whales' backs over rock ridges, then plunges into the trough. The roar drowns all other sounds. The best I can do is to avert the highest humps, the deepest drops, and the biggest holes below. The water is a glassy, deceptively smooth mass that plunges six feet. At the bottom a huge wave curls back. "Straighten out!" Leif screams. My hands freeze at the oars.

We enter the spillway at a 45 degree angle, not straight on so that our bow could blast through. The curved wall of water slams the side, slapping me to the rubber floor. River pours over the tube. The boat slows and one side rises. Chris grips the rope. No time to high side. I grab for the oars.

At last the hole spits us out. The spare oar has broken loose, but there's no way to reach it. Worse, we're full of water, dead weight in the rampant current. The river sweeps us along, encumbered by several hundred pounds of ballast, toward the worst rapid on the run: Skull. While Leif and Chris bail madly, I heave my body into the oars. Arms whirling like a windmill, Betty waves us in. Hard strokes haul us across the current.

In no hurry to see Skull Rapid, I pick my way over black

rocks too hot to touch. Sheer walls constrict the entire river into something resembling a narrow flume below a floodgate. At the bottom a deep hole churns into angry turbulence. To the right lurks the Room of Doom, where an eddy swirls within a sheer-walled cavity. Beyond the millrace looms a jagged wall—the Rock of Shock—where the current piles up. Skull derives its name from a sheep that, trapped in the Room, left its bleached skull there as a warning.

My belly tightens and even Betty looks pale. Though the roar limits communication, her face says: "If you almost went over in Funnel, you'll flip here for sure." This is no place for mistakes. Boaters who've ended up in the Room of Doom have clung to rocks for hours, even days, until someone could reach them with ropes. It does look possible to cheat Skull to the left, where the drop is less drastic and the waves are less daunting. This plan should keep us away from both the Room and the wall. But getting set up for this shot poses a problem. It's tough to work against the current to get positioned. And it'll be tricky to shoot straight out from the bank, cut across the current to the precise spot, spin, and then slice left down the chute. We're threading a course between Scylla and Charybdis—between a wicked jetty and a whirlpool that devours sailors.

I lean as hard as I can into my new, untested, nine-foot oars. Feet wedged under the thwart, faces furrowed, Leif and Chris paddle us into the swift current. This is the spot. Now . . . down the chute. Barking commands, I heave power strokes away from the maelstrom. The paddlers dig in with equal fury. We skirt the turbulence but graze the jetty, which spins our bow downstream. I bury my left oar, registering the strain in my shoulder socket. Leif and Chris swivel back to face the bow as we skirt the mound of river along the

Rock of Shock.

We whoop to celebrate before we look back to see how the larger boat will fare. To our surprise, hot-dogging kayakers come shooting right into the seething madness, helmets nearly submerged in the spray. One kayak flips. I cringe as the current sucks it upside-down right toward the Rock. At the last minute the kayaker rights himself, grinning, oblivious to his close call. Moments later Betty's boat enters the flume, gaining momentum to power through the turbulence. Legs fly up as her raft hits a big wave, then fists thrust skyward as she emerges right side up. When she nears, the two crews pitch buckets of river at each other, releasing tensions.

Sock It To Me, Bowling Alley, and other rapids extend the rollicking ride, but the fury is over. Spectacular red cliffs embellish the last five miles to Rose Ranch, where the others end their run. As we say our goodbyes, I thank them for a rip-roaring ride. Though the unrelenting wind tempts me to stop here until it does relent, the allure of a rarely-boated stretch pulls me onward, alone but not lonely.

Where the canyon widens again, the river broadens as its current slows to where it's no help against the wind. One channel heads south, another veers west around an island that promises some shelter, though not enough. Once more the wind forces me off the river, leaving me with almost thirty miles to cover on the last day.

Bleached grasses droop down the cutbank, their roots hanging like dusty wires. I scramble up a beaver slide, rope in hand. Because it's not grazed, this island looks pristine. Tall native grasses wave in thick clumps that bend beneath my sandals. Its leaves swept closed, a squawbush flexes in the wind. Green caterpillars have stripped some of its leaves, exposing the dark-red berries. These early fruits are old friends. On many a hot day their tang has stimulated saliva to slake thirst.

When the wind abates, the late-day light shimmers on the river, allowing the eye to trace the flow of gusts as they roughen its surface. Dust in the air mutes the colors, graying and purpling the red rocks in the distance. Southwest of here the volcanic La Sal peaks rise like pyramids; even on their northern slopes the snow is almost gone. While the daylight lingers, cliff swallows swoop in their characteristic long arcs, each ending in a bank and climb.

The dreamy twilight turns golden, then grainy like an old sepia photograph. Soon the waxy moon spotlights the river's boils and swirls. Nearly invisible, a great blue heron stalks in the waning twilight. When I make the slightest move, the heron flows into flight, its wings making the slightest rustle. It trails raucous "graks," alarm calls well known to river runners. Young herons must invest more time fishing than older, experienced ones. I'll never know whether this was a young bird, possibly working overtime because the afternoon winds made it difficult to fly or to see prey, or an old one, its eyesight still keen, that's learned to stalk by moonlight.

Another day dawns with brilliant sunshine. Tan and green spiders from the grasses, their webs sparkling with beads of dew, have commandeered the *Canyon Wren*. Given the keen sensitivity of spiders to the slightest vibration, these stowaways freeze as soon as I touch the rope. Too often our lives become spider webs of precautions, attempts to register every threat through its subtle vibrations.

As I flick the spiders out, one by one, they bounce on the river's surface and scoot for the nearest cattail. Spiders rank among the most efficient land predators anywhere, but on this island they neither compete with, nor get eaten by, lizards, which hunt both bugs and spiders. Islands without lizards often harbor a hundred times more spiders than those with them.

For the next twelve miles, the broad brown river meanders through open ranch country. Buttes of Morrison shales and cliffs of red Kayenta and Wingate sandstone rise beyond the verdant banks where two-rut roads wander down to the river's edge. The chugging of pumps heralds ranches.

Two turkey vultures, an unusual sight in these parts, roost in a dead cottonwood while another gorges on a fish. These are large black birds, smaller only than condors and eagles. When they feast on larger animals, these scavengers plunge their nearly featherless heads deep into the carcass. As their bare heads get exposed to the sun, heat and dryness inhibit the growth of bacteria or parasites picked up during feeding.

Vultures circle, often in flocks, riding the thermals for hours until one bird locates a carcass. Though vultures possess an acute sense of smell, they can't sniff out carrion from great heights. If they don't find food, they may practice "piracy," or kleptoparisitism, whereby they force a nesting great blue heron to regurgitate food to feed their own young. In part because they're trusting of humans, who often dread them as symbols of death, and in part because they're ugly, turkey vultures once declined to a point where they appeared on the Audubon Society's Blue List.

Before long Cisco Wash enters on the right. Bisected by Interstate 70, the Cisco Desert looks like barren hills of clay. Despite limited forage, these hundred thousand or so acres are favored by white-tailed prairie dogs. But decades of overgrazing by sheep plus several dry years have diminished the already minimal forage plants. One result was a dramatic die-off of prairie dogs, those ground-dwelling squirrels that sit up and wag their tails, wrestle like puppies, and kiss with their lips parted. Cuteness is not the issue, however, for these rodents provide essential food for other wild critters.

Like magpies and ravens, prairie dogs have suffered the consequences of human ignorance. Somewhere along the dusty trail, ranchers declared war on them and enlisted government exterminators. The resulting onslaught—enlisting 132,000 killers in 1920 alone—has included bombing, gassing, drowning, target shooting, germ warfare, poisoned food, and even a huge vacuum device that sucks the dogs from their burrows, still alive. Our tax dollars at work.

Rancher folklore holds that these rodents compete with livestock for forage, but research suggests that prairie dogs actually improve range by aerating and fertilizing the soil to promote more mixed plant communities. One Forest Service study, for instance, concludes that both "wildlife and domestic livestock preferentially feed on these prairie dog colonies." Yet the killing continues, led not only by ranchers but by "sportsmen" who blow the dogs to pieces with telescopic rifles. South of here in Naturita, Colorado, each year hundreds of hunters converge for a slaughter that seems to vent anger and indulge a need for power. In the West especially this mentality may be a holdover from the pioneer attitude that anything wild should be subdued. For similar reasons many ranchers kill other "varmints"—including weasels, badgers, skunks, foxes, and jackrabbits.

Prairie dogs are the primary prey of endangered black-footed ferrets, a weasel-like mammal now proliferating in captivity. Reflecting current theory in population biology, Jean Akens and Dave May observe in *Canyon Legacy* that when vegetarians decline, so do carnivores: "One naturally thinks of predators as 'controlling' the populations of prey species." In fact, simplistically, the opposite is nearer the truth: "the availability of prey species controls the number of predators." This assumes, of course, that people are not also killing off the predators.

By the early 1980s biologists feared that black-footed ferrets had become extinct—victims of overgrazing, drought, plague, canine distemper, and the extermination of prairie dogs. But eighteen survivors were captured alive and bred in captivity. Under the supervision of the U.S. Fish and Wildlife Service, these ferrets multiplied like people. Their reintroduction poses problems, however, for few large prairie dog towns remain and many of these experience recurrent epidemics. If the Bureau of Land Management cooperates by reducing domestic sheep, the Cisco Desert could become a prime habitat for the ferrets' comeback in the wild.

With oars tucked behind my knees, the current spins the *Canyon Wren* like an air mattress bobbing on the Gulf Stream. After days on the river, I grow close to it, adopting its muddy and ruddy hues. The oars stir the total fluidity; unguided by thought, I make contact with water and earth and air. Whereas confronting the dark trench of rapids calls for concentration on every rock, wave, and eddy, drifting occasions a different engagement, an intimacy that allows me to row by feel.

Discovering how much I can rely on myself, how I can row a river alone, frees me when I reenter society. When I know that I can handle the unforeseen, I feel more free to take risks. Most of all, greater intimacy with nature seems to enlarge the possibilities for intimacy with people.

Colder, freer of flotsam yet loaded with silt, the Dolores River enters on the left. Nearby Dewey Spring attracted early settlers whose log cabins still stand beneath ancient cottonwoods. Dewey Bridge, though bypassed by a contemporary

cement structure, still spans the river. For decades this picturesque one-lane suspension bridge was the only crossing between Fruita, Colorado, and Moab, Utah—a distance of one
hundred and ten river miles.

This final stretch is the most scenic. As it streams toward
the heart of the Colorado Plateau, the river winds through
Professor Valley. Here it presents a world of strong colors:
brown river, white beaches, green banks, and rich red outcroppings such as Fisher Towers and Castle Rock, all set
beneath a deep blue sky. For some reason, the usual winds
don't blow. It's hot but the rapids toward the end deliver
welcome spray. Remarkably, given the development in
nearby Moab, the Colorado is secluded today.

Lured downstream, I reluctantly stroke into the shallows
at Sandy Beach takeout above the boundary to Arches National Park. As I unstrap the frame, other boaters exude a
hearty spirit as they tear down their rigs. One crusty river
runner, her sun-streaked hair swinging under a cap, celebrates
with stories. She tells of a commercial trip that flipped in
Funnel Falls and spent the night in the gorge until a search
and rescue team could reach the hapless boaters.

Come to think of it, we were lucky. If we'd flipped in the
half-mile inner gorge, we probably couldn't have pulled out
of that seven-mile-an-hour millrace before we'd have had to
abandon the raft above Skull. Even if we'd been able to
extract an overturned boat from the channel, the three of us
probably couldn't have righted it in deep water. Neither the
BLM brochure nor the river guidebook mentions these possibilities.

As I swig beer in the shade, a couple of seasoned boaters
from Moab comment on dams. Since 1884, when the first
major dam was completed in California, thirty-seven thou-

sand dams have been built west of the Mississippi. Over a century, Western water experts Donald Worster and Marc Reisner both note, the West has lived a paradox. It considers itself a place of freedom and democracy but has become a hydraulic society dependent on government water projects for the privileged few. Christian traditionalist C. S. Lewis observed that "what we call man's power over nature turns out to be a power exercised by some men over other men with nature as its instrument."

Sinking water tables, half-filled reservoirs, and rivers reduced to trickles all suggest that the West faces severe water shortages not just for its burgeoning human population but for its flora and fauna as well. Climate changes could further intensify shortages. Worst of all, the West's water laws allow far-away entities like Las Vegas to claim any water that is not demonstrably serving a "beneficial" function. Beneficial, of course, has traditionally been defined strictly in economic terms with no regard for recreation, let alone for other species. If we're to maintain the already minimum flows not just for river running but for sport fish, game animals, vanishing species, and native plants, this definition will need substantial revision.

The romantic in me wants to rhapsodize as the Colorado rolls and roils, now a dozen times more muddy than the Mississippi, carrying the continent to the Pacific. But the realist knows that it drops its load in Lake Powell, where polluted sediments settle on industrial junk. Barely a trickle crosses the Mexican border, where the once-mighty river hasn't reached the sea since the early 1960s.

Despite such depletion, even degradation, the Colorado Plateau remains an enormously enchanting region, a truly magical place that takes hold of the soul.

Rhapsody

The Swerve in the Moonlight

My tires whir like June bugs as I bike Zion's curvy red road. While cool air streams by my face, a breeze animates fall foliage aglow with stored sunshine. Pink maples and yellow box elders linger, weathering the frosts, their fluorescence soft against the dusky walls. The deer are already browsing as I coast by, bugs in my nose. As the canyon walls tighten, the domed rims lean in. Far above, the sunset gilds the Great White Throne.

Soon I enter the Temple of Sinawava where, like waterspouts after a shower, cool air pours from clifftops. Small fish, probably dace or minnows, silver the shallows. Below the Narrows, bubbles accompany crickets that chirp a twilight melody. Bats swoop and flutter amid gnats that bob in the dusk. As a beaver gnaws its trunk, a golden-hearted cottonwood glows its last for the day.

High up on the rim, a still-invisible moon blanches slickrock that gleams like snow. As the moonbeams creep downward, silvered ponderosa pines far up the cliffs become frosted blue spruces. The light becomes palpable, the rock only a chimera.

The Paiutes called Zion *Ioo Goon*, "the quiver," a place to "go out the way you came in." They didn't stay after dark but tonight, though the Great White Throne now stands in shadow, the spirits seem supremely benign.

On the way out I sail through the liquid moonlight. With the traffic gone for the night, my senses open wide in amazement. A sandstone symphony plays music for my eyes, melodies I hear with my heart.

Antlers burst from the darkness. I lunge for the handle-bars, grip the brakes hard, and veer. Hooves click and clunk on pavement as the buck bounds past, oblivious of me. With a timely swerve, man and beast share the way.

As the poet Rumi observes, "There are many ways to kiss the ground."